Pocket Book of Technical Writing for Engineers and Scientists

Leo Finkelstein, Jr.
Wright State University

Boston   Burr Ridge, IL   Dubuque, IA   Madison, WI
New York   San Francisco   St. Louis
Bangkok   Bogotá   Caracas   Lisbon   London   Madrid   Mexico City
Milan   New Delhi   Seoul   Singapore   Sydney   Taipei   Toronto

Mcgraw-Hill Higher Education

A Division of The **McGraw-Hill** Companies

Vice president/Editor-in-Chief: *Kevin T. Kane*
Publisher: *Thomas Casson*
Executive editor: *Eric Munson*
Editorial coordinator: *Michael Jones*
Senior marketing manager: *John T. Wannemacher*
Project manager: *Kimberly D. Hooker*
Production supervisor: *Debra Benson*
Designer: *Jennifer Hollingsworth*
Cover illustration: © *Nishan Akguilian / SIS*
Compositor: *Lachina Publishing Services*
Typeface: *10/12 Century Schoolbook*
Printer: *R. R. Donnelley & Son Company*

Library of Congress Cataloging-in-Publication Data

Finkelstein, Leo, 1946–
    Pocket book of technical writing for engineers and
    scientists / Leo Finkelstein, Jr.
        p. cm.
    ISBN 0-07-237080-7 (pbk.)
    1. Technical writing--Handbooks, manuals, etc. I. Title.

T11 .F53 2000
808'.0666--dc21   99-044806

http://www.mhhe.com

Pocket Book of Technical Writing for Engineers and Scientists

McGraw-Hill's *Best—Basic Engineering Series and Tools*

Bertoline, *Introduction to Graphics Communications for Engineers*

Burghardt, *Introduction to Engineering Design and Problem Solving*

Chapman, *Introduction to Fortran 90/95*

Donaldson, *The Engineering Student Survival Guide*

Eide, et al., *Introduction to Engineering Design*

Eide, et al., *Introduction to Engineering Problem Solving*

Eisenberg, *A Beginner's Guide to Technical Communication*

Finkelstein, *Pocket Book of Technical Writing for Engineers and Scientists*

Gottfried, *Spreadsheet Tools for Engineers: Excel '97 Version*

Gottfried, *Spreadsheet Tools for Engineers: Excel 2000 Version*

Greenlaw and Hepp, *Introduction to the Internet for Engineers*

Palm, *Introduction to MATLAB for Engineers*

Pritchard, *Mathcad: A Tool for Engineering Problem Solving*

Schinzinger/Martin: *Introduction to Engineering Ethics*

Smith, *Project Management and Teamwork*

Tan and D'Orazio, *C Programming for Engineering and Computer Science*

This book is dedicated to the memory of my father, who taught me all the important things I needed to know.

About the Author

Leo Finkelstein, Jr., received his bachelor's degree from the University of North Carolina at Chapel Hill in 1968; his master's from the University of Tennessee at Knoxville in 1969; and his Ph.D. from Rensselaer Polytechnic Institute at Troy, New York, in 1978. He is currently Lecturer and Director of Technical Communication for the College of Engineering and Computer Science, Wright State University, Dayton, Ohio. As an associate professor, he directed the technical writing program at the U.S. Air Force Academy while also serving as an adjunct associate professor for the University of Colorado at Colorado Springs. Prior to that, he wrote, produced, and directed technical films in Southern California and commanded a combat documentation photographic unit in Southeast Asia during the Vietnam War. In addition, his military service includes experience in both space and logistics systems. He holds FCC commercial and amateur radio licenses, has a black belt in tae kwon do, and is a senior member of the Society for Technical Communication.

Preface

Just who would write a book like this? I would. I am Leo Finkelstein, Jr., and I have been writing about technical subjects and teaching technical writing for more years than I want to remember. I have also been doing a lot of other things.

I suppose it all started when I was a kid growing up in Asheville, North Carolina. I routinely terrorized the neighbors with powerful ham radio transmissions and various high-voltage experiments. The radio transmissions were just a few harmonics away from everyone's favorite television channel, and the experiments would be considered a serious environmental threat today. I had to learn to write technically about what I was doing, especially when responding to violation notices from the Federal Communications Commission, not to mention all the complaints from the attorney who lived across the street.

In high school I was a combo guy for the local radio station, meaning that I did two adult jobs (disc jockey and transmitter engineer) for the price of one kid. I found that the station's management did not care how bad I sounded as long as I worked cheaply enough and kept the station on the air. I spent my undergraduate summers working in television in Washington, D.C., as a camera operator on a nationally televised wrestling show from "Capital Arena." I found it strange how the director always seemed to know in advance when a huge brawl was about to take place. He would order the cameras into position minutes before the big fight "spontaneously" erupted.

Years later, after graduating with a master's degree from the University of Tennessee at Knoxville, I entered the Air Force as a motion picture writer/producer/director. I produced and directed technical films on a variety of subjects, ranging from electromagnetic spectrum management and geodetic analysis to civil engineering and aircraft maintenance. Since few artistic writers wanted anything to do with these technical scripts, I also routinely wrote the films I produced and directed. Somehow this qualified me to command a combat photographic unit for a year in Southeast Asia during the Vietnam War, where I became heavily involved in combat aerial documentation, precision recording photography, and something new—bureaucratic administrative writing.

Later, after getting my Ph.D. at Rensselaer Polytechnic Institute, I became an associate professor at the U.S. Air Force Academy. In that role I directed the academy's technical writing program, which included about 1,600 students and 11 instructors each year. I also taught for the University of Colorado at Colorado Springs as an adjunct associate professor in the evenings, and I spent a couple of years at Air Force Space Command working advanced concepts and writing all kinds of reports.

I retired from the Air Force after 20 years, and the next day, I joined the College of Engineering and Computer Science at Wright State University, Dayton, Ohio. I am currently Lecturer and Director of Technical Communication for the college. This book represents my approach to teaching technical writing. It is a straightforward, laid-back method that works but does not take itself too seriously. The first part of the book discusses the component skills necessary for technical writing, including technical definition, mechanism description, and process description. The second part shows how to produce the most common types of technical documents used in engineering and the sciences. I have also included a chapter on writing resumes and application letters for technical positions, along with useful tips for job interviewing.

In the last part of the book I address grammar and style; I have also included practical discussions of documentation, visuals, technical briefings, electronic publishing, technical writing ethics, and abstracts.

Technical writing is absolutely essential to the effective functioning of all engineering and science disciplines. I believe that fluency in technical writing will enhance the career of anyone working in these fields. In that regard, I sincerely hope all of you can put this book to good use.

Dr. Leo Finkelstein, Jr.
College of Engineering and Computer Science
Wright State University
Dayton, Ohio 45435

Acknowledgments

Acknowledging those who helped me with this book is a daunting task because there were so many who did so much. In particular, however, I thank my wife and best friend, Phyllis A. Finkelstein, whose nonstop encouragement and superb proofreading were nothing short of miraculous. Of course, a lot of her friends say that her putting up with me for more than 30 years is also nothing short of miraculous.

I thank my friend, Dr. William G. Dwyer, a senior analyst, retired Air Force colonel, current technical writing teacher, and onetime Shakespeare scholar, for his encouragement and proofreading of the entire manuscript. I thank my friends and colleagues, Dr. Thomas A. Sudkamp, Professor of Computer Science and Engineering, and Dr. Fred D. Garber, Associate Professor of Electrical Engineering, for their service "above and beyond the call of duty" as metaphorical backboards against which I have bounced a plethora of crazy ideas.

For a more youthful perspective on this project, I thank my son and computer science student, Stephen B. Finkelstein, for reviewing my chapters and "telling dad exactly what he thought." And I thank my former student and current friend, Sharon E. Liebel, a practicing biomedical engineer and ergonomics expert, for providing her unique perspective and insight.

Finally, my sincere appreciation to all those who provided formal reviews for this book. In particular, I acknowledge the contributions of Dr. Kenneth J. Breeding, Professor of Electrical Engineering, The Ohio State University; and Ms. Suzanne Karberg, Communication Specialist, Civil Engineering, Purdue University. I also appreciate the thorough copyediting of the manuscript by Ms. Meg McDonald.

Contents

Pocket Book of Technical Writing for Engineers and Scientists

1

Introduction

Technical writing is a fundamental skill for virtually anyone working in science and engineering—and that includes a broader range of people than just scientists and engineers. Most science and engineering activities produce technical reports either on paper or in electronic form. Government and commercial activities involving research, development, finance, and commerce rely on precisely written documents to communicate complex information to a wide range of audiences and for many purposes. Technical writing is the means by which these documents are produced.

What Is Technical Writing?

To define what technical writing is, it might be useful to first clarify what it *is not.* Technical writing is not what one does out in the meadow under an elm tree; that is *creative writing.* Creative writing is, by and large, a pleasant activity. Technical writing, on the other hand, is tough, hard work. Technical writing is not what most people commonly do for fun.

It is not difficult to imagine a learned, creative person sipping fine wine, eating fancy cheese, and watching the sun set. This person sounds like a creative writer, who with paper and pen in hand might reflect on the nature of time as the last rays of sunlight slowly disappear over the horizon. Our creative writer might even come up with a definition something like this:

Time is a river flowing from nowhere through which everything and everyone move forward to meet their fate.

A creative, sensitive person sees an inspiring sunset, is moved to words, and writes the "river from nowhere" definition. Obviously, this approach is metaphorical. But what if someone inspired by that magnificent sunset were to write this definition of time?

Time is a convention of measurement based on the microwave spectral line emitted by Cesium atoms with an atomic weight of 133 and an integral frequency of 9,192,631,770 Hz.

Perhaps not. What normal person would be emotionally moved by a sunset and then come up with microwave spectral lines and Cesium 133? This might be a pretty strange person. The second definition of time is technical. It is designed to be objective, direct, and precisely defined in an empirical manner. Consequently, it lacks the emotional impact of the first definition because, as technical writing, it avoids the use of rich metaphors and figures of speech, substituting instead precise, empirical data.

The difference between the two definitions shows the fundamental distinction between technical writing and creative writing (and all the other kinds of writing that fall in between). Technical writing is precise, objective, direct, and clearly defined.

Abstraction   In linguistic terms, technical writing is writing that displays a relatively low level of abstraction. To clarify the concept of *abstraction,* consider the ladder of abstraction in Figure 1.1.[1] Here various levels of abstraction are being used to refer to a

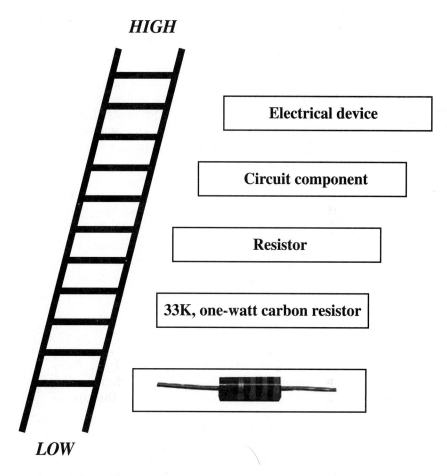

Figure 1.1
Abstraction ladder

33,000-ohm, one-watt carbon resistor. The lowest level of abstraction would be the resistor itself. However, we normally do not paste resistors into technical documents; so to be precise, we have to substitute something else, like a photograph. That is why a photograph of the resistor has been placed at the bottom of the abstraction ladder—because that is the most concrete way available for referring to it in a document. This precision, of course, assumes that the audience knows what a resistor looks like. If not, the photograph can be pretty abstract, and for that matter, so can the actual resistor.

The next most concrete way of describing the resistor using words alone is to precisely label it—as we do when we give our newborn children distinctive names and IRS taxpayer identification numbers. In this case, the resistor is labeled a "33K, one-watt carbon resistor."

Moving up the ladder of abstraction, the references become increasingly imprecise. At the next level, "resistor" could mean any circuit component that impedes the flow of current. At the subsequent level, "circuit component" could mean any electrical device in the circuit, such as a capacitor, inductor, diode, or, in this case, a resistor. Finally, at the top of the abstraction ladder, "electrical device" could be anything that is electrical in nature, which could be anything from a fluorescent light, to a computer keyboard, to a stereo system. Of course, "electrical device" also refers to that 33K, one-watt carbon resistor.

The point is that as one moves up the ladder of abstraction, the symbols become less precise and increasingly vague. In effect, this increased abstraction gives the reader greater freedom to interpret meaning as he or she wants, and not necessarily as the writer intended. In creative writing, that is probably good; in technical writing, that is *always* bad. The goal of technical writing is to eliminate abstraction. Simply speaking, successful technical writing restricts the reader's freedom of interpretation so that only one meaning can be concluded—the meaning intended by the writer.

A given reader might interpret that river flowing out of nowhere as any number of things based on his or her prior experience and emotional makeup. (I know that I learned to water ski on a river flowing out of nowhere—I believe it was called the Potomac!) Unlike the river from nowhere, Cesium 133 is Cesium 133, and 9,192,631,770 Hz is 9,192,631,770 Hz, no matter what the reader's prior experience or emotional makeup. What makes the technical definition of time technical, then, is that it effectively restricts the reader's ability to abstract various meanings.

The potential for abstraction can never be totally eliminated, even by the best technical writing. The process of human perception presupposes some abstraction. Who knows—the reader might have flunked out of Chemistry 101 as an undergraduate 10 years ago because he could not find Cesium in the periodic table of elements. Perhaps flunking chemistry caused him to lose his scholarship, which, in turn, led to his leaving the university. He was subsequently dumped by his girlfriend, the valedictorian of their high school and the one great love of his life. In despair, he became a bum. She, ironically, became a Nobel Prize–winning chemist. So for this reader, the technical definition of time might have a connotative, emotionally charged meaning far apart from its precise, denotative function. It could happen; but in reality, it is certainly not likely.

What sets technical writing apart, then, is its precision. How it achieves this precision is, in fact, the art and craft of technical writing—an activity that involves definition and description; data and analysis; photographs, diagrams, and charts; and often specialized language. The goal of technical writing is *not* to be creative or interesting; it is *not* to employ rich imagery or powerful metaphors. The goal of technical writing, first and foremost, is to communicate complex information clearly and precisely for the audience and the purpose at hand. Clarity and precision are the overriding goals for any technical writer, and audience and purpose are the primary considerations for achieving those goals.

Audience and Purpose   The measure of how well a technical writer has written something comes down to two things: (1) how well the reader understands, precisely, the writer's intended meaning; and (2) how well that understanding fulfills the intended purpose or the need at hand. Consequently, technical writing must be geared directly to its audience and purpose. Remember, there is always some specific requirement for technical writing: a scientist needs to write a proposal for a grant; a programmer needs to document a software package prior to distribution; or a lab chief needs to write a feasibility report for selecting new equipment.

To further relate specifically to the purpose and situation at hand; it also has to relate specifically to the reader, or audience, that will be using the document. In other words, the writer must consider the potential reader's knowledge, skill level, and specialization; and the writer has to fully respond to the needs of the reader in terms of the requirements of the situation.

To further clarify these concepts, let us briefly define a requirement and audience. Suppose that person watching the sunset who was writing about spectral line emissions and time has a problem with his social life. The problem is that he does not have one. By random chance, let us say he has finally met a young lady who has captured his heart. She, of course, scarcely knows he exists. He has decided to rectify the situation by writing her a note articulating his love for her. Coming up with the right words is the requirement, and she is the audience.

The man has spent the day reading psychology journals on the concept of love and now believes he has the requisite knowledge to write about the subject. So that night, he pens a note to the woman of his dreams:

> Whenever I look into your eyes, I know that, from
> my perspective, I share with you a strong, interper-
> sonal passion or enthusiasm statistically related at
> .05 or better to increased levels of self-disclosing
> behavior.

This writer may have the right idea, but he is
doing the wrong kind of writing for his purpose
and audience.

The same thing would be true of a social scientist preparing a scholarly paper on the occur-
rence patterns of erotic love among middle-aged
men and women. Suppose this scientist writes
the following:

> Love is the bond that holds humanity together and
> the rapier that rips it apart.

Certainly he or she could make this point, but
given the purpose of the academic conference and
the nature of his scholarly audience, this scientist probably would have to reword the passage
to read something like this:

> In terms of interhuman social structures, love is an
> emotional quality that can exhibit adhesive or divi-
> sive properties or functions in response to outside
> stimuli.

This definition is technical writing; as such, it is
more suited for a group of social scientists
researching interpersonal communication patterns of subjects in experimental and control
groups.

Report Writing and Audience

In technical writing, the audience and purpose
are almost always well defined in advance—usually by the boss. The reason is that technical
writing is basically report writing, meaning that

it is writing commissioned by someone else for a specific purpose and audience. Normally, reports aim to share objective information with an interested, educated audience. Technical reports are simply reports that share this information in a very precise way.

So, finally, to answer the question posed at the beginning of this chapter, *technical writing* can be defined as follows:

- *Technical writing deals with technical information.* As the previous examples show, using technical writing for nontechnical purposes and situations is a bad idea. Technical writing is designed to deal with technical subjects and nothing else.
- *Technical writing relies heavily on visuals.* As shown in the abstraction ladder of Figure 1.1, a photograph or diagram may be the least abstract, and therefore the most precise, way to communicate something to a technical audience. Visuals— whether they are equations, photographs, tables, graphs, drawings, or charts—are powerful ways to provide a large amount of information effectively and efficiently; however, they almost always require interpretation or explanation in the text of the report, especially when the audience may not be familiar with them.
- *Technical writing uses numerical data to precisely describe quantity and direction.* In many cases, mathematical equations and values provide the real substance of a technical report.
- *Technical writing is accurate and well documented.* Generalized, unsupported assertions have no place in a technical paper. Conclusions, recommendations, and judgments are always based on clearly presented evidence or established expertise, and technical writing is *always* technically correct.

- *Technical writing is grammatically and stylistically correct.* This kind of correctness is far more than simply a matter of personal and corporate pride. Grammar and style often go to the heart of the author's credibility. In other words, if you write like an idiot, the reader may well perceive you as an idiot and your organization as a collection of idiots. Whether this perception is true is irrelevant. Where technical writing is concerned, perception is often reality.

The following pages provide a straightforward, easy-to-follow tutorial on how to do technical writing. These pages go to great lengths not to take themselves too seriously. The goal here is not to pontificate but rather simply to get to the point, providing the necessary information in a way that is interesting and understandable.

The next three chapters describe the basic components of definition and description because these constitute the foundation of technical writing. The next several chapters specifically focus on the most common types of technical writing, including proposals; progress reports; feasibility and recommendation studies; instructions and manuals; laboratory and project reports; research reports; and resumes, cover letters, and job interviews. The rest of the book provides an easy-to-understand treatment of grammar and style and looks at the most common errors in technical reports today. It provides a no-nonsense guide to technical documentation, including when and how to use it. The latter chapters also show how to design and use visuals in technical documents and address the differences between traditional and electronic publishing. A straightforward discussion on ethics in technical writing is also included, as well as a chapter on writing abstracts and executive summaries.

Notes   1. The idea of an abstraction ladder was borrowed from S. I. Hayakawa, *Language in Thought and Action* (New York: Harcourt, Brace and Company, 1949), p. 160.

2

Technical Definition

Virtually any kind of technical writing includes one or more technical definitions. Consequently, a technical writer must be able to define terms, irrespective of whether these terms refer to mechanisms or processes.

In technical writing, definition is the process by which one assigns a precise meaning to a term. To define a term, it must be placed into a classification, then differentiated from other terms in that same classification. Technical definitions are relatively easy to write, except for some pitfalls that will be addressed later. The format for a technical definition is straightforward and works like this:

What Is a Technical Definition?

Term = <u>Classification</u> + Differentiation

For example, if a writer were to define the stall condition an airplane experiences when it loses lift, he or she could start with the term *stall,* then add a classification, <u>flight condition</u>, and then differentiate it from all other flight conditions—in this case, by a stall's unique characteristics. The definition might read something like this:

A *stall* is a <u>flight condition</u> where the lift produced becomes less than the weight of the airplane, and the airplane stops flying.

That seems simple enough; but what happens when a term, like *stall*, has multiple definitions in many contexts? In such cases it may be necessary to add a qualifier in front of the definition statement to supply the necessary context. The qualifier is important when the general context for a definition needs to be established up front. If the context is known or is obvious, a qualifier is unnecessary. For example, in an aeronautics study on aircraft wing design, the context of *stall* is obvious. It is clear that *stall* in this case has to do more with the loss of lift than with, say, a single compartment for an animal in a barn.

When a context is needed, the format for the definition would be

$$(\text{Qualifier } +) \; \textit{Term} \; = \\ \underline{\text{Classification}} + \text{Differentiation}$$

By the way, the parentheses, italic font, and underscore have been added only for clarity.

Look at Figure 2.1, which provides three definitions of the same term in different contexts. Note how the term *stall* has three totally different meanings depending on the context, and how each definition begins with a qualifier that makes the context clear from the start.

In the first example, *stall* refers to what happens when an airplane does not go fast enough to stay in the air. Pilots routinely stall their airplanes right above the runway when landing them, in which case stalling is good. Sometimes they stall them inadvertently in much more precarious situations, in which case stalling is bad.

In the second example, *stall* refers to a car that has suddenly stopped running. Normally, this condition happens only in the middle of heavy traffic, in bad weather, and with a critical appointment on the route.

The third example of *stall* relates to social dating behavior. (This is the one with which I have

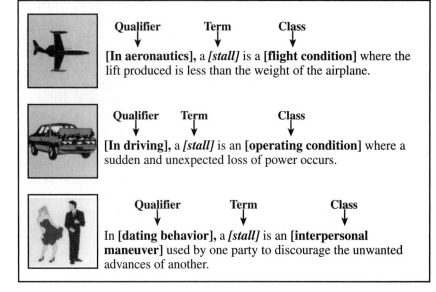

Qualifier   Term   Class

[In aeronautics], a *[stall]* is a [flight condition] where the lift produced is less than the weight of the airplane.

Qualifier   Term   Class

[In driving], a *[stall]* is an [operating condition] where a sudden and unexpected loss of power occurs.

Qualifier   Term   Class

In [dating behavior], a *[stall]* is an [interpersonal maneuver] used by one party to discourage the unwanted advances of another.

Figure 2.1
Multiple contexts and qualifiers

had the most experience, especially back in my undergraduate years. As I remember, the typical "good night kiss" stall was for my date to fill her mouth with bubble gum and start chewing.)

Classifications and Classes

Often the most difficult part of writing a technical definition is coming up with the proper classification for the term. The class should be a general category in which the term fits, but it cannot be too general. For example, consider that 33K, one-watt carbon resistor from the abstraction ladder in Chapter 1. In the following sentence, the term is defined using the very generic classification *device.*

The *33K, one-watt carbon resistor* is a <u>device</u> that impedes the flow of electrical current.

The problem with this classification is that *device* could mean all kinds of different things, most of which have nothing to do with circuit components; consequently, its inclusion does not really help specify the meaning of the term. By changing *device* to *circuit component,* however, the meaning can be narrowed considerably for the reader even before it is differentiated.

One trick for classifying a term is to build an abstraction ladder for the term, then move up one or two "rungs" above the term. In the original abstraction ladder, the movement was as shown in Figure 2.2. In this case, moving up one rung—from the term "33K, one-watt carbon resistor" to the term "resistor"—is not an option because the classification would be derived from the original term. That would yield the following circular definition:

The *33K, one-watt carbon resistor* is a <u>resistor</u> that impedes the flow of electrical current.

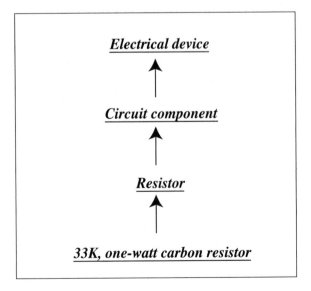

Figure 2.2
Abstraction ladder

Because the 33K, one-watt resistor is obviously a resistor, this classification does not help define the term. In fact, it contributes nothing other than useless circularity. If possible, try to find a classification that is not derived from the term. In this case, the easiest solution is to just move up the abstraction ladder one more rung. *Circuit component* helps to define the term and has no circularity with the term.

Differentiation

The next step in defining the term is to differentiate it from all the other members of the class. Differentiation involves narrowing the meaning of the term to just one possibility within the class. Clearly, it would be easier to narrow the class of "circuit components" to a particular resistor than it would be to narrow "device" to a particular resistor. There are all kinds of devices in the world and relatively few circuit components.

The class "circuit components," however, still contains many possibilities: capacitors, diodes, switches, potentiometers, inductors, transistors, and IC chips, for example. In this case, a good approach is to focus on the function of a resistor, which is to impede the flow of electric current, and to use that function to differentiate the class. Doing so yields the following definition:

The *33K, one-watt resistor* is a circuit component that impedes the flow of electric current.

Avoiding Mistakes

In writing technical definitions, it is easy to do something bad. For example, define the common computer term *hard drive*. It is probably a good idea to qualify the definition unless the context of computing is obvious. (There are other kinds of hard drives, such as one through the Mojave Desert without air conditioning.)

So how about this definition?

> In computing, a *hard drive* is an input/output device for the nonvolatile storage and retrieval of data.

Think about this one for a minute. If the reader does not know what a "hard drive" is, what is the likelihood that the reader is going to know what an "input/output device" is, much less the meaning of "nonvolatile storage and retrieval"? As mentioned earlier, an important rule of writing definitions is never to define a term with the same term. Sometimes, however, an even more important rule is never to define a term with another term that itself needs defining. Consider the reader's knowledge and skill level. In this case, the term was defined with terms that need even more defining.

Extensions   As with many rules, there are times when you may have no choice but to violate the principle to achieve the goal. Sometimes a simple term that the audience will understand is just not available. In fact, sometimes the only thing to do is to define the term with undefined terms, then extend the definition immediately by explaining the undefined terms. For example:

> In computing, a *hard drive* is an input/output device for the nonvolatile storage and retrieval of data. *Input/output devices* move information into and out of the computer. The computer writes (or *outputs*) data to the hard drive and reads (or *inputs*) data from the hard drive. *Nonvolatile storage* means that when data is written to the hard drive, it remains there even when the power is turned off.

The three sentences added in this definition are called *extensions*. Extensions are sentences that follow up a definition statement with additional information the reader needs. Often a single-sentence definition will not be adequate. In such cases, you will have to extend the definition in a way that communicates to the audience and serves the purpose at hand.

Extensions can take many forms. In the example just given, the extensions further define the terms used to define the original term. But extensions can do far more than provide further definition. Here are a few of the most common types of extensions you can use, along with examples that relate to our hard drive:

Common Extensions

- Use *comparison and contrast* when you need to show differences or similarities.

  Hard drives differ from optical drives in that hard drives store data as magnetic patterns read by a read/write head, whereas optical drives store data as pits and lands read by a laser.

- Use *classification* when you need to organize information into categories.

  Hard drives are categorized by the size of their storage capacity, the speed at which data can be accessed, and the type of computer interface they employ.

- Use *cause and effect* when you need to demonstrate why something happens or when you need to trace results.

  Hard drives write information by passing electric currents representing data through a coil wound on a highly permeable material called a *head*. The resulting electromagnetic pulses cause magnetic patterns to be recorded on the surface medium of a spinning disk.

- Use *process* when you need to list the steps of a procedure.

  To select Drive D in Windows, first double-click the "My Computer" icon on the desktop, and then, in the resulting window, double-click the "Drive D" icon.

- Use *exemplification* when you need to give real or analogous examples.

  Hard drives have evolved rapidly in terms of storage capacity and speed. For example, the Finkel-DRIVE 1000 provides 1,000 gigabytes of storage with access times less than 10 nanoseconds. (I made this up, of course.)

- Use *etymology* to show the linguistic genesis of the term.

  The term *magnetic disk drive* comes from the ancient Latin *magneticus,* meaning "mysterious flux"; *diskae,* meaning "shaped like a plate"; and *drivum,* meaning "to rotate with torsion." (If you believe any of this is true, I have a Finkel-DRIVE you can buy at a bargain.)

Remember, the choice of which extensions you use should be governed by the purpose of your paper and the knowledge requirements of your audience. In other words, if you are writing a set of instructions for a novice operator on how to use a hard drive, extensions that involve the basic use of the hard drive would be more appropriate than extensions that address the theoretical basis for the hard drive's operation.

Required Imprecision   In some situations you will need to trade off desired precision in your definition to achieve the required level of communication. At times it is foolish to attempt to achieve expert-level precision with an uninformed audience.

Consider the following two definitions of a *black hole*—the astrophysical phenomenon that is supposed to exist somewhere in space.

> In astrophysics, a *black hole* is a set of events from which it was not possible to escape to a large distance. A black hole gets its name from its boundary, called an event horizon, which is formed by the paths in space-time of rays of light that just fail to get away, hovering instead forever on the edge and, consequently, moving on paths parallel to or away from one another.

> In astrophysics, a *black hole* is a collapsed neutron star whose gravity is so great that even light cannot escape. Although fusion reactions within this collapsed star still may emit brilliant rays of light, viewed from the outside, the black hole appears to be a totally dark void in space.

The first definition functions at the expert level. For theoretical physicists, it provides a precise and accurate description of a black hole and thus is appropriate for their needs. But for the less informed, reading it represents a mind-twisting experience that, in many cases, can leave readers more confused regarding the term than they were before they read it.

The second definition functions at the level of the average reader. It is not nearly as precise or accurate as the first definition, but it communicates the basic gist of what constitutes a black hole in space.

The problem is that to be precisely correct and absolutely accurate, you would have to function at a level where you cannot effectively communicate to the average reader. If you were a technical writer writing for the average reader, you would have to make a tough decision here: either be absolutely correct and communicate less effectively, or be less than absolutely correct and communicate more effectively. If your goal is effective communication, then your decision is obvious.

A Word about Defining Specifications and Standards

Defining specifications and standards is a specialized activity. Such documents can take many different forms, depending on what areas of engineering and science are involved and whether the documents are designed to meet commercial, industrial, or government standards. Writing these kinds of definitions is far more complex than simply defining terms.

Specification documents precisely state particulars, including requirements, designs, implementations, and testing. In engineering and science, specifications normally involve goods and services being developed under some type of contractual obligation. The specification precisely defines the quality of work and performance standards required by the contract.

Specifications are also what technical standards are made of. Standards are accepted or established methods, measures, or designs for accomplishing specific tasks. Many standards exist for everything from data transfer protocols and cable connectors to air conditioning coolants and drinking water.[1] Specifications used in standards are detailed and exacting. The following excerpt from the IEEE 1394 Open Host Controller Interface Specification is a good example:

1394 requires certain 1394 bus management resource registers be accessible only via "quadlet read" and "quadlet lock" (compare-and-swap_ transactions), otherwise ack_type_error shall be sent. Those special bus management resource registers are implemented internal to the 1394 Open Host

Controller to allow atomic compare-and-swap access from either the host system or from the 1394 bus.[2]

If the specification is required under U.S. government contract, it must contain certain information about the goods and services involved and the various standards that apply. Government specifications often require the following:

- Precise definitions and descriptions of the scope of the project.
- Any documentation the contractor must furnish, along with the formats for those documents.
- Specific performance characteristics of any required product, along with necessary testing, including procedures and equipment, to verify that the goods or services meet the specified requirement.
- Exact descriptions of the deliverables of the contract, including all goods and services, and the dates and times by which these products will be provided.
- Contractor notes, records, and other research and production materials.

Writing specifications is a demanding task normally accomplished by experienced engineers, scientists, and project managers with a solid knowledge of all applicable standards. Standards are often developed by committees composed of legal, managerial, and technical experts.

- Have I fully analyzed the purpose of my report, and do I understand the skill and knowledge level of the audience?
- Have I defined the term by first classifying it in a way that adds precision and understanding for my audience and serves my purpose?

Definition Checklist

- Have I differentiated this classification to distinguish this term from other members of its class?
- Have I determined whether the context is clear and, if not, whether it is critical to the definition? If the context is unclear and critical, have I used a qualifier before the term?
- Have I avoided defining a term with the same term?
- Have I avoided using terms that themselves need to be defined? If not, have I explained these terms?
- Have I chosen extensions to my definitions that are appropriate for my audience and purpose?
- Have I compromised my fundamental purpose (communicating with my reader) by including inappropriate or irrelevant information or precision?

Exercise   Read each of the following definitions and try to determine the context, audience level, accuracy, and purpose for which they were written.

- A *resistor* is a small electronic part that reduces the amount of electricity flowing through a circuit.
- A *resistor* is a circuit component that converts electrical energy into thermal energy and, in the process, determines the current produced by a given difference of potential.
- *Resonance* is a systemic condition where small amplitudes of a periodic agent produce large amplitudes of oscillation or vibration.
- *Resonance* is a natural means of amplification that makes a musician's horn sound louder.
- *Ionization* is the electrostatic process by which a neutral atom or molecule loses or gains electrons, thereby acquiring a net charge.
- *Ionization* is the phenomenon that creates lightning in thunderstorms.

- *Ergonomics* is the field of study by which we make machines easier to use.
- *Ergonomics* is the systematic consideration of physical, psychological, and social characteristics of human beings in the design of tools and equipment, the workplace, and the job itself.

Notes

1. For an updated, comprehensive listing of standards, see "CFS Standards Document Library on the World Wide Web." Internet: http://www-library.itsi.disa.mil/by_org.html, March 23, 1999. Some of the more commonly used standards on this Web site include American National Standards Institute (ANSI) http://www-library.itsi.disa.mil/org/ansi_std.html; Department of Defense Standards (DOD-STD) http://www-library.itsi.disa.mil/org/dod_std.html; Institute of Electrical and Electronic Engineers (IEEE) http://www-library.itsi.disa.mil/org/ieee_std.html; International Organization for Standards (ISO) http://www-library.itsi.disa.mil/org/iso_std.html; Military Standard (MIL-STD below 2045) http://www-library.itsi.disa.mil/org/mil_stdb.html; Military Standard (MIL-STD 2045 and up) http://www-library.itsi.disa.mil/org/mil_std.html; and Telecommunications Industry/Electronic Industries Association (TIA/EIA) http://www-library.itsi.disa.mil/org/tia_eia.html.

2. Apple Computer, Inc., Compaq Computer Corporation, Intel Corporation, Microsoft Corporation, National Semiconductor Corporation, Sun Microsystems, Inc., and Texas Instruments, Inc., "1394 Open Host Controller Interface Specification," Release 1.00, p. 38. Internet: http://1394ohci-l@austin.ibm.com.

3

Description of a Mechanism

Technology involves mechanisms, and being able to describe these mechanisms precisely and accurately in a way the reader can understand is perhaps the most essential component skill of writing technical reports. This skill is particularly important for those producing documents involving specifications and instructions.

What Is a Mechanism Description?

Mechanism descriptions are precise portrayals of things. *Things* are material devices with two or more parts that function together to do something. Mechanisms can range in complexity from circuit components and mechanical fasteners, to supercomputers and space shuttles, to swing sets a parent has to put together on Christmas morning.

In Outline 3.1, notice that the primary focus in writing a mechanism description is on the physical characteristics or attributes of a device and its parts. These documents are built around precise descriptions of size, shape, color, finish, texture, and material. Such descriptions also normally include figures, diagrams, or photographs that directly support the word discussion.

Example of a Mechanism Description

To show how Outline 3.1 works, we will use it to describe a relatively simple mechanism: the 33K, one-watt carbon resistor discussed in the first two chapters. The resistor is a relatively simple

---

Outline 3.1   Description of a Mechanism

Introduction
- Define the mechanism with a technical definition (see Chapter 2) and add extensions to discuss any theory or principles necessary for the reader to understand what you are saying. *Always make sure you add only what the reader needs for the purpose at hand.* If the reader does not need any theory or operating principles, do not provide any.
- Describe the mechanism's overall function or purpose.
- Describe the mechanism's overall appearance in terms of its shape, color, material, finish, texture, and size.
- List the mechanism's parts *in the order in which you plan to describe them.*

Discussion
- Part #1
  - Define the first part with a technical definition, adding extensions as needed to deal with theory or operating principles.
  - Describe the part's overall function or purpose.
  - Describe the part's shape, color, material, finish, texture, and size using precise measures and descriptors. Also be sure to use figures, diagrams, and photographs as necessary.
  - Transition from this part to the next part.
- Parts #2–n
  - For each remaining part, repeat the pattern of defining, describing, and transitioning established for Part #1.

Conclusion
- Briefly summarize the mechanism's function and relist the parts described.
- Give a sense of finality to the paper.

---

mechanism; most mechanisms are much more complex. Assume, for the purpose of illustration, that this description is for the average technical reader who requires only a general description of the resistor.

Introduction

Following the outline, first introduce the mechanism with a technical definition and extensions that describe its overall function and purpose:

> The 33K, one-watt carbon resistor is a circuit component that impedes the flow of electrical current.

Next, use extensions to discuss any theory or operating principles necessary for the reader to understand the description. Be careful here; think about *what* you are doing, *for whom* you are doing it, and *why* you are doing it. Do not lose sight of the reader's knowledge and skill level or forget the purpose of the report. The goal is not to show how smart you are, but rather to communicate the information. So, when discussing theory in a mechanism description, a good rule is to do what is necessary, but only what is necessary.

For example, consider this theoretical discussion:

> The resistor impedes the movement of free electrons, thereby generating a thermal response depending on temperature, cross section, and length of the resistive element. The resulting resistance is measured in ohms, where a resistor has one ohm of resistance when an applied electromotive force of one volt causes a current of one ampere to flow. In addition, the square of the current flowing in amps, times the resistance in ohms, determines the power dissipated in watts. This particular resistor can safely and continuously dissipate one watt of electrical energy as heat.

This discussion might be fine for readers with more specific purposes, but not for the average person who needs only a general description. Remember: the goal here is to write a general, informative mechanism description. The purpose is *not* to provide a description of how a resistor works or to show how to measure its effect in a circuit. In addition, if this theoretical discussion were to be used for the intended audience, several additional concepts (such as current, voltage, and power) would have to be defined and discussed as well.

Given the reader and purpose, the following two sentences might provide a simplified theoretical discussion that is more appropriate:

> The resistor impedes the flow of current by converting a portion of the electrical energy flowing through it into thermal energy, or heat. This particular resistor can safely convert one watt of electrical energy into heat.

Notice that this description does not get into Ohm's Law. (It also does not explain that if I placed 1,000,000 volts across this resistor, it would draw 30.3 amperes of current, resulting in the dissipation of more than 30 million watts of power, thereby generating an inferno that would destroy my house and probably the entire neighborhood.) This information is not relevant to *describing the mechanism*. It might be relevant to describing the operation of the mechanism (or in answering the civil suits brought by the neighbors), but that is not the purpose here.

The next step in the introduction is to describe, in general terms, the mechanism's overall appearance: its shape, color, material, finish, texture, and size. In this case, one could describe the device as follows:

> The 33K, one-watt carbon resistor looks like a small cylinder with wire leads extending from each end. The cylinder's surface is composed of smooth, brown plastic with a shiny finish. Four equally spaced color bands (three orange, one gold) circumscribe the cylinder starting at one end.

Finally, to complete the introduction, list the mechanism's parts in the order in which they will be described. This listing organizes the remainder of the mechanism description. Consequently, the decision regarding the order in which to list the mechanism's parts is not trivial. It effectively determines the structure for the rest of the mechanism description.

There are two ways to order the parts: spatially and functionally. Using a spatial organization, you can move from left to right, or top to bottom, or inside out, or outside in. Using a functional approach, you can order the parts in terms of how the parts function with one another. Functionally, for example, you could start with the *leads,* which connect the circuit to the *carbon element,* which is protected by the *casing,* and around which the *color bands* are painted to indicate resistance and tolerance values. Using this approach, the final sentence of the introduction might read as follows:

> The resistor consists of the following parts: two wire leads, the carbon element, the casing, and the color bands.

Or using, for example, an inside-out spatial approach, the final sentence of the introduction might read this way:

> The resistor consists of the following parts: the carbon element, the wire leads, the casing, and the color bands.

Both approaches are fine as long as they are logical and make sense to the reader. In any case, once the parts have been listed, move on to the discussion section, where detailed descriptions of the parts are provided.

Discussion

The discussion section of a mechanism description precisely describes, in the necessary detail, each part of the mechanism. It always follows the organizational pattern established by the listing of parts at the end of the introduction. For example, using inside-out spatial ordering, the discussion section will have four subsections: the carbon element, the leads, the casing, and the color bands.

The discussion section is laid out in a relatively simple manner: create a subsection for each part, then begin by describing the first part in the first subsection, and so forth.

Carbon Element

First, define the part with a technical definition:

> The carbon element is the capsule of resistive material that converts electrical energy into heat.

Second, provide an extension that describes the part's function and gives any needed theory:

> The carbon element serves as the primary active component of the resistor and provides the necessary 33,000 ohms of resistance. The element functions by blocking, to some degree, the flow of free electrons passing through it. The energy released by these blocked free electrons is then dissipated in the form of heat.

Next, describe the part's shape, color, material, finish, texture, and size, using precise measures and descriptors. Be sure to include necessary visuals such as diagrams and photographs. Three visuals will be added to this example later in this chapter, but for now, concentrate on describing the part's physical attributes:

> The carbon element is cylindrically shaped and is 2.4 cm long with a diameter of .31 cm. It is composed of finely ground carbon particles mixed with a ceramic binding compound. The element is gray with a dull, matte finish.

Then, transition to the next part by showing how this part relates to it:

The carbon element is electrically connected to the leads.

Now handle the rest of the parts in similar fashion.

Leads

Define the part:

The leads are two conductive wires connected to opposite ends of the carbon element.

Describe the leads' function and provide any needed theory:

The leads actually have two functions. First, they provide electrical connectivity from the carbon element to the circuit; and second, they provide a mechanical means of mounting and supporting the resistor in the circuit environment.

Describe the part in detail:

The leads, which have a dull, silver color and smooth texture, are composed of tinned copper wire. Each lead is four cm long and is 20 gauge in thickness.

Transition to the next part.

The leads connect to the carbon element through the ends of the casing.

Casing

Define the part:

The casing is a cylindrical enclosure that surrounds the carbon element.

Describe its function and provide any needed theory:

The function of the casing is twofold. First, it physically protects and insulates the carbon element from the outside environment. Second, it provides the heat-exchanging medium needed to dissipate the thermal energy generated by the carbon element.

Describe the part in detail:

The casing is a brown, plastic cylinder that is 2.5 cm long, with a .312-cm inside diameter and a .52-cm outside diameter. It snugly fits over the carbon element.

And transition to the next part:

The outside of the casing is circumscribed by four color bands.

Color Bands

Define the part:

The color bands are visual indicators that describe the resistance and tolerance of the resistor.

Describe the bands' function and provide needed theory:

Using the standard color code for commercial, four-band resistors, and starting with the band at the edge of the cylinder, the first three bands represent the value of the resistor in ohms. The fourth band indicates the tolerance or accuracy of the resistor.

Describe the part in detail:

Each color band is .1 cm wide and is circumscribed around the outside of the casing and parallel to the edge of the casing. Each color band is smooth and shiny. The first color band, which starts flush with one end of the casing, is orange and represents a value of 3. The second color band, which starts .1

cm away from the inside edge of the first color band, is orange and also represents a value of 3. The third color band, which starts .1 cm away from the inside edge of the second color band, is orange and represents a multiplier of 1000. The fourth color band, which starts .1 cm away from the inside edge of the third color band, is gold and represents a tolerance of 5 percent.

Now that all of the parts have been described, it is time to add a conclusion.

Conclusion

The final section of the mechanism description serves two purposes: it summarizes the description of the mechanism, and it provides a sense of finality to the document.

First, briefly summarize the mechanism's function and relist its parts:

The 33K, one-watt carbon resistor is a circuit component that impedes the flow of electrical current through the use of a carbon element. The resistor is made up of four parts: the carbon element, which impedes the flow of current by converting a portion of the electrical energy applied into heat; the wire leads, which electrically connect the element to the circuit and support the resistor mechanically; the casing, which encloses and insulates the element and dissipates heat from it; and the color bands, which indicate the resistance and tolerance of the device.

Now give a sense of finality to the paper. Include a sentence that by tone and content indicates to the reader that the mechanism description is complete:

Together, these parts form one of the most commonly used circuit components in electronic systems today.

This final sentence tells the reader not to look for anything else because the document is ending; it is a courtesy to the reader.

Visuals and Mechanism Descriptions   When designing visuals such as diagrams, figures, and photographs in support of a mechanism description, be sure to include only information that directly relates to the mechanism description and is specifically keyed to the description provided in the text. Do not include visuals that do not match the mechanism description in subject matter or terminology. Also, avoid visuals that include too much or too little complexity for the level of discussion in the paper.

The following is the complete mechanism description. Note that three visuals have been added to enhance the description, including two diagrams and a photograph. Pay particular attention to these factors:

1. These visuals specifically relate to the text discussion, and they are referred to in the text before actually being used. The reader should never encounter a visual and wonder why it is there or what it is for. Always tell your reader specifically when to look at a visual, and try to do so before the point in the paper where the visual actually appears.

2. Each visual has an assigned sequence number and name—such as Figure 3.3 Cutaway view. These labels reference precisely the visuals that are included. By the way, it is a good idea to use compound numbers—to show both the section or chapter number and the sequence number. You will notice in the following example that the first figure is labeled Figure 3.1. The "3" indicates that the visual occurs in the third chapter of this book, and the "1" refers to the sequence in which

the figure occurs. Numbering visuals in this way allows adding or deleting visuals in one section without having to renumber all the visuals in subsequent sections. Also, if possible, run separate sets of sequence numbers for each type of visual. For example, figures should have their own set, as should tables and photographs.

3. Each visual must be included for a purpose. In the following description, Figure 3.1 provides a visual overview of the device and labels its main parts using a cutaway for the internal carbon element. Figure 3.2 provides an X-ray view of the resistor that shows the various physical dimensions. Figure 3.3 shows a cutaway of the entire device.

Description of a 33K, One-Watt Carbon Resistor

Putting It All Together

Introduction

The 33K, one-watt carbon resistor is a circuit component that impedes the flow of electrical current. The resistor impedes the flow of current by converting a portion of the electrical energy flowing through it into thermal energy, or heat. This particular resistor can safely convert one watt of electrical energy into heat.

The 33K, one-watt carbon resistor looks like a small cylinder with wire leads extending from each end. The casing's surface is composed of smooth, brown plastic with a shiny finish. Four equally spaced color bands (three orange, one gold) circumscribe the cylinder starting at one end.

The resistor consists of the following parts: the carbon element, the wire leads, the casing, and the color bands (see Figure 3.1).

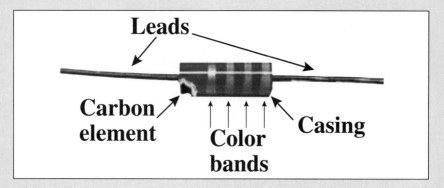

Figure 3.1
Parts of the resistor

Discussion

For the following discussion, refer to Figure 3.2 for an X-ray view of the resistor's parts, along with its physical dimensions, and Figure 3.3 for a cutaway view of an actual resistor.

Carbon Element

The *carbon element* is the capsule of resistive material that converts electrical energy into heat. The carbon element serves as the primary active component of the resistor by providing the necessary 33,000 ohms of resistance. The element functions by blocking, to some degree, the flow of free electrons passing through it. The energy released by these blocked free electrons is then dissipated in the form of heat.

The carbon element is cylindrically shaped and is 2.4 cm long with a diameter of .31 cm. It is composed of finely ground carbon particles mixed with a ceramic binding compound. The element is gray with a dull, matte finish. The carbon element is electrically connected to the leads.

Leads

The *leads* are two conductive wires connected to opposite ends of the carbon element. The leads actually have two functions. First, they provide

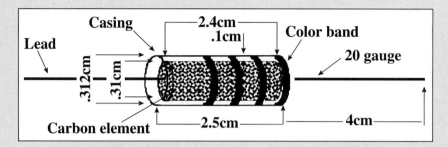

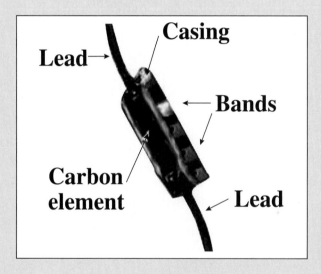

Figure 3.2
X-ray view of
resistor with
dimensions

Figure 3.3
Cutaway view

electrical connectivity from the carbon element to the circuit; and second, they provide a mechanical means of mounting and supporting the resistor in the circuit environment. The leads, which have a dull, silver color and smooth texture, are composed of tinned copper wire. Each lead is 4 cm long and is 20 gauge in thickness. The leads connect to the carbon element through the ends of the casing.

Casing

The *casing* is a cylindrical enclosure that surrounds the carbon element. The function of the casing is twofold. First, it physically protects and electrically insulates the carbon element from the outside environment. Second, it provides the heat-exchanging medium needed to dissipate the thermal energy generated by the carbon element. The casing is a brown, plastic cylinder that is 2.5 cm long, with a .312-cm inside diameter and a .52-cm outside diameter. It snugly fits over the carbon element. The outside of the casing is circumscribed by four color bands.

Color Bands

The *color bands* are visual indicators that describe the resistance and tolerance of the resistor. Using the standard color code for commercial, four-band resistors, and starting with the band at the edge of the cylinder, the first three bands represent the value of the resistor in ohms. The fourth band indicates the tolerance or accuracy of the resistor. Each color band is .1 cm wide and is circumscribed around the outside of the casing and parallel to the edge of the casing. Each color band is smooth and shiny.

The first color band, which starts flush with one end of the casing, is orange and represents a value of 3. The second color band, which starts .1 cm away from the inside edge of the first color band, is orange and also represents a value of 3. The third color band, which starts .1 cm away from the inside edge of the second color band, is orange and represents a multiplier of 1000. The fourth color band, which starts .1 cm away from the inside edge of the third color band, is gold and represents a tolerance of 5 percent.

Conclusion

The 33K, one-watt carbon resistor is a circuit component that impedes the flow of electrical current through the use of a carbon element. The resistor is made up of four parts: the carbon element, which impedes the flow of current by converting a portion of the electrical energy applied into heat; the wire leads, which electrically connect the element to the circuit and support the resistor mechanically; the casing, which encloses and insulates the element and dissipates heat from it; and the color bands, which indicate the resistance and tolerance of the device. Together, these parts form one of the most commonly used circuit components in electronic systems today.

Mechanism Description Checklist

- Have I defined the mechanism, and have I extended this definition with any theory or principles necessary for my reader's understanding? Have I included theory or principles that either are not needed or are above or below the level of my reader?
- Have I described the mechanism's overall function or purpose?
- Have I described the overall appearance of the mechanism in terms of its shape, color, material, finish, texture, and size?
- Have I listed the main parts in the introduction?
- Have I defined each of these parts?
- Have I described each part's function or purpose?
- Have I discussed the needed theory or operating principles for each part?
- Have I precisely described each part's physical structure?
- Have I provided transitions from one part to the next?
- Have I concluded by summarizing the mechanism's function and parts?
- Have I given a sense of finality to the description?

Exercise   Apply the checklist just given to the following mechanism description of the FinkelBOAT 688 Attack Submarine. What problems exist? Are any parts missing? Do the definitions work? Are any of them circular? Do some require additional definitions or extensions? Can you determine the intended audience and the purpose of the description? Pay particular attention not only to the size and complexity of this mechanism, but also the purpose, presentation, and quality aspects of the description. What about the visual? Is the level of complexity appropriate for this description? Is the visual properly marked? Does it appear in the correct place? What purpose does it serve? How well integrated is the visual into the text's description? Also, what sections provide precise technical writing, and what sections feature abstract, imprecise concepts?

Mechanism Description of a FinkelBOAT 688 Attack Submarine

Introduction

The FinkelBOAT 688 submarine is a nuclear-powered, attack, undersea watercraft characterized by high speed, ultraquiet operation, high cost, and an awesome array of conventional and special weapons. The submarine, which is used primarily for peaceful undersea research, has a top submerged speed of 44 knots and carries both UGM-109 cruise missiles and Mk-48 torpedoes. The FinkelBOAT 688 consists of a main tube, two hemispheric caps, a propeller, and a fairwater.

Discussion
Main Tube
The *main tube* is a claustrophobic cylinder of pure YM-31 high-tensile titanium that provides

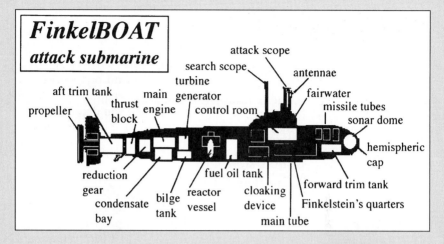

FinkelBOAT
attack submarine

attack scope
search scope
turbine
antennae
aft trim tank
main generator
fairwater
propeller
thrust engine
missile tubes
block
control room
sonar dome
reduction gear
fuel oil tank
hemispheric cap
condensate bay
bilge tank
reactor vessel
cloaking device
forward trim tank
Finkelstein's quarters
main tube

Figure 3.4
FinkelBOAT
submarine

the body of the boat. The main tube is 75 feet
long, six inches thick, and has an inside diameter
of 33 feet. The main tube is coated with an ane-
choic/decoupling coating designed to defeat
active sonars and contain internal noise. At each
end of the main tube are hemispheric caps.

Hemispheric Caps

The *hemispheric caps* are cones of pure YM-31
high-tensile titanium that enclose both the stern
and bow ends of the main tube, thereby completing the substructure of the boat. The caps are
tapered and welded to the main tube's barrel sections. The boat's propeller is located directly
behind the stern hemispheric cap.

Propeller

The *propeller* is a seven-bladed propeller that
provides high thrust with little or no cavitation.
The propeller, which is made of a special bronze
alloy, is 30 feet in diameter and can silently
move 500 acre feet of water per second at flank
speed. On the top of the substructure, and 55
feet forward toward the bow hemispheric cap, is
the fairwater.

Fairwater

The *fairwater* is the raised, enclosed observation post of the submarine that supports periscopes and communications antennae and that also provides a modest bridge area. Often called a sail or conning tower, the fairwater, which is 25 feet high, 15 feet long, and 10 feet wide, is also composed of pure YM-31 high-tensile titanium.

Conclusion

As the figure shows, the FinkelBOAT has lots of parts that work together to make a submarine.

4

Description of a Process

Process descriptions are similar to mechanism descriptions. But whereas mechanism descriptions describe the physical attributes of a mechanism, process descriptions describe the steps of a process.

Processes are important in technical writing because much of what needs to be communicated involves unfolding events or actions. In fact, process descriptions are key elements of many technical reports. Process descriptions can include objective, third-person portrayals of events that do not directly involve the reader—like explaining to your teenager how a modern car operates. They can also be second-person descriptions for human involvement that provide specific instructions so that the reader can perform or accomplish the process—like showing your teenager how to drive the car. Normally, giving specific instructions is a more challenging task.

What Is a Process Description?

Process descriptions are precise portrayals of events occurring over time that lead to some outcome. Process descriptions describe either the steps in the operation of a mechanism or the steps of a conceptual process. Unlike mechanism descriptions, process descriptions of a mechanism in operation focus less on the physical attributes of the mechanism and more on a mechanism's function and how the parts work together. For example, a description of a car's four-cycle, internal combustion engine in operation would not focus on the parts (such as the pistons, rods, cylinders, valves, and spark plugs); rather, the description would focus on the steps of the engine's operation (such as intake stroke, compression stroke, ignition stroke, and exhaust stroke). By the same token, a description of how to drive the car might include steps such as fastening the seat belt, starting the engine, engaging the foot brake, releasing the parking brake, and putting the transmission in gear.

As with mechanisms in operation, conceptual process descriptions discuss the steps of a process, but these steps do not involve a physical mechanism. For example, a process description for operations research might well include a decision tree; however, this tree is not the kind you have to plant, water, and fertilize. In fact, it is not a thing at all. It exists only conceptually in the imagination.

Whether a process involves something physical is irrelevant. The organization of a process description is basically the same, whether it describes a mechanism in operation or a conceptual process. Additionally, the general organization of the process description is the same whether it provides a third-person account of what happens in the process or second-person instructions on how to accomplish the process.

Outline of a Process Description

Notice that the two process description outlines here are similar. Outline 4.1 lays out the pattern for describing the operation of a mechanism, such as an air conditioning system that includes compression, condensation, expansion, and evaporation as its steps. Outline 4.2 provides a similar pattern for describing a conceptual process, such as a sort algorithm that includes the iterative steps of evaluation, identification, and insertion.

Outline 4.1 Process Description of a Mechanism in Operation

Introduction
- Define the mechanism with a technical definition (see Chapter 2) and add extensions to discuss any theory or principles necessary for the reader to understand what you are saying. Make sure you include only what the reader needs for your purpose.
- Describe the purpose, function, and operation of the mechanism.
- List the major steps of the mechanism's operation.

Discussion
- Step #1
  - Define the step with a logical definition and discuss it in reasonable depth.
  - Describe the equipment, material, or concepts involved in this step.
  - Describe what happens during this step.
  - Show the relationship between this step and the next step with a transition statement.
- Steps #2–n
  - For each remaining step, repeat the pattern established for Step #1. The final step discussed may not have a transition, unless it cycles back to the first or subsequent step.

Conclusion
- Briefly summarize the mechanism's function and the major steps of its operation.
- Give a sense of finality to the paper.

The following descriptions illustrate the use of these outlines to describe processes. The first example describes the operation of a mechanism. The second example describes a purely conceptual process that does not involve any type of mechanism.

Process Description Examples

Description of a Mechanism in Operation

To provide an example for using Outline 4.1, this chapter will describe the process of the operation of a mechanism—in this case, a notional air conditioning system. We will add visuals later once

> ## Outline 4.2 Description of a Conceptual Process
>
> Introduction
> - Define the process with a technical definition (see Chapter 2) and add extensions to discuss any theory or principles necessary for the reader to understand what you are saying. Make sure you include only what the reader needs for your purpose.
> - Describe the purpose and function of the process.
> - List the major steps of the process.
>
> Discussion
> - Step #1
>   - Define the step with a logical definition and discuss it in reasonable depth.
>   - Describe the purpose and function of this step.
>   - Describe what happens during this step.
>   - Show the relationship between this step and the next step with a transition statement.
> - Steps #2–n
>   - For each remaining step, repeat the pattern established for Step #1. The final step discussed may not have a transition, unless it cycles back to the first or subsequent step.
>
> Conclusion
> - Briefly summarize the function and the major steps of the process.
> - Give a sense of finality to the paper.

the process has been developed. The audience for this description is a general, technical reader who does not have specific expertise in air conditioning, fluid mechanics, or thermodynamics.

Introduction

Begin by introducing the mechanism with a logical definition, then add extensions as necessary to provide needed theory or principles:

An *air conditioner* is a mechanical device used to refrigerate a controlled environment. The air conditioner accomplishes this refrigeration by transferring heat within the environment to an area outside the environment.

Next describe the function and operation of the mechanism:

> The air conditioner transfers heat from one area to another using a fluid refrigerant. This refrigerant is pumped through both the controlled environment and the outside area. At the same time, the refrigerant is cycled at strategic points between liquid and vaporous states. This change in state provides the means for transferring thermal energy.

You may want to list the primary operational parts of the mechanism, especially if they relate directly to the process steps. In any case, be sure to list the major steps of the mechanism's operation in the order in which they will be described. Choosing this order may not be easy. Ideally, the steps will follow a logical timeline from the start to the finish of the process; however, in some cases, such as transactional, iterative, and branching processes, that may not be possible. In such cases you will have to describe the process steps in the way that is clearest for your reader and the purpose at hand.

> The air conditioner is built around four major components, including the compressor, condenser coil, expansion valve, and evaporator coil. The operation of the air conditioner relates directly to these parts and includes the following steps: compression, condensation, expansion, and evaporation.

Discussion

In the discussion section, treat each step as a separate subsection. Define each step of the process, then extend these definitions as necessary to address the equipment, material, or concepts involved. Be sure to tell your reader clearly what happens in each step, and provide linking transitions between the steps.

Compression.   Discuss compression in reasonable depth for the audience and purpose at hand. First define *compression,* then deal with the equipment and concepts involved. Describe what happens during compression, and then show the relationship with the next step, condensation:

> *Compression* is a fluid dynamics process in which a given volume of refrigerant vapor is forced to occupy a smaller volume of space. Compression occurs when the compressor forces hot refrigerant vapor under pressure into the compression chamber. The chamber is composed of a cylinder/valve arrangement. The piston draws refrigerant into the cylinder through an intake valve, the intake valve closes, and the piston pushes up into the cylinder, compressing the refrigerant vapor. The vapor then exits through an exhaust valve and enters the condenser coil, where the condensation step occurs.

Condensation.   Describe condensation in much the same way as you dealt with compression. Start with a technical definition of *condensation,* then extend this definition as necessary to describe the parts of the mechanism involved and the concepts and theory the reader may need. Then describe what actually happens in this step of the process, and transition to the third step, expansion:

> *Condensation* is a fluid dynamics process in which the hot refrigerant vapors, pumped from the compressor to the condenser coil, cool and change into a liquid. As this change occurs, the heat of condensation is released from the refrigerant in the condenser coil, heating the coil. This heat, in turn, is drawn from the coil by a fan, which passes relatively cooler air across the coil, picking up the heat

and venting it outside. The refrigerant liquid then flows through a closed loop to the expander valve, where rapid expansion occurs.

Expansion.   Treat this third step in the same manner as the first two. First define *expansion,* then extend the definition to describe the parts of the mechanism involved and to provide needed theory and concepts. Explain what actually happens during this step, and provide a transition to the fourth step, evaporation:

> *Expansion* is a fluid dynamics process in which the condensed liquid refrigerant, under relatively high pressure from the compressor, is forced through an expansion valve into an area of substantially lower pressure. The expansion valve acts as a nozzle, constricting and then accelerating the liquid refrigerant until it passes through a threshold where the constriction is removed and rapid expansion occurs. At this point the expanding refrigerant enters the evaporator coil, where it will cool and change state again.

Evaporation.   Finally, define the last step, *evaporation,* extend the definition, and describe what happens as in earlier steps. However, because this process is cyclical and repetitive, the transition from this last step should link back to the first step:

> *Evaporation* is a fluid dynamics process in which the rapidly expanding refrigerant liquid changes into a vapor. As the liquid enters the evaporator coil, it also enters an area of substantially lower pressure. As a result, it vaporizes, and in the process absorbs heat. Air in the controlled environment is circulated across this coil, which, in turn, absorbs heat from the air. The fan distributes the resulting cooler air throughout the controlled environment. The refrigerant vapor in the evaporator coil is then drawn back through the closed loop to the compressor, at which point the entire cycle repeats.

Conclusion

The last part of the process description summarizes the process and provides a sense of finality as a courtesy to the reader:

> The operation of an air conditioner involves four steps. First, the compressor pumps refrigerant under pressure into the condenser coil. Here it is liquefied, giving up heat that is removed by a fan circulating air over the condenser coil. The liquid refrigerant moves through a closed loop, through an expansion valve, and into the lower-pressure evaporator coil. Here the refrigerant changes into a vapor, absorbing heat from air passing over the evaporator coil. This air then cools the controlled environment, while the refrigerant is drawn back into the compressor. At this point the cycle is complete, and the process repeats.

Visuals and Process Descriptions

Designing visuals for process descriptions follows the same rules and has the same considerations as designing visuals for mechanism descriptions—except for one significant difference. Process visuals usually need to show something happening and, as such, tend not to be static representations of something seen in a "slice of time." Often you can show movement through space and time with a technique as simple as a well-placed arrow. At other times you may need a coherent series of visuals to demonstrate your point. Additionally, well-integrated captions within the visual can help explain it.

| Putting It All Together | Description of the Operation of an Air Conditioner |

Introduction

An *air conditioner* is a mechanical device used to refrigerate a controlled environment. The air conditioner accomplishes this refrigeration by transferring heat within the environment to an area outside the environment. The air conditioner transfers heat from one area to another using a fluid refrigerant. This refrigerant is pumped through both the controlled environment and the outside area. At the same time, the refrigerant is cycled at strategic points between liquid and vaporous states. This change in state provides the means for transferring thermal energy.

The air conditioner is built around four major components, including the compressor, condenser coil, expansion valve, and evaporator coil. The operation of the air conditioner relates directly to these parts and includes the following steps: compression, condensation, expansion, and evaporation. See Figure 4.1.

Figure 4.1
Air conditioning system

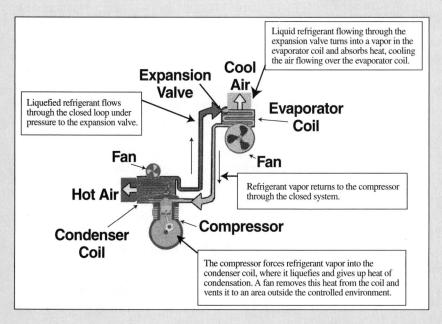

Liquid refrigerant flowing through the expansion valve turns into a vapor in the evaporator coil and absorbs heat, cooling the air flowing over the evaporator coil.

Liquefied refrigerant flows through the closed loop under pressure to the expansion valve.

Refrigerant vapor returns to the compressor through the closed system.

The compressor forces refrigerant vapor into the condenser coil, where it liquefies and gives up heat of condensation. A fan removes this heat from the coil and vents it to an area outside the controlled environment.

Compression

Compression is a fluid dynamics process in which a given volume of refrigerant vapor is forced to occupy a smaller volume of space. Compression occurs when the compressor forces hot refrigerant vapor under pressure into the compression chamber. The chamber is composed of a cylinder/valve arrangement. The piston draws refrigerant into the cylinder through an intake valve, the intake valve closes, and the piston pushes up into the cylinder, compressing the refrigerant vapor. The vapor then exits through an exhaust valve and enters the condenser coil, where the condensation step occurs.

Condensation

Condensation is a fluid dynamics process in which the hot refrigerant vapors, pumped from the compressor to the condenser coil, cool and change into a liquid. As this change occurs, the heat of condensation is released from the refrigerant in the condenser coil, heating the coil. This heat, in turn, is drawn from the coil by a fan, which passes relatively cooler air across the coil, picking up the heat and venting it outside. The refrigerant liquid then flows through a closed loop to the expander valve, where rapid expansion occurs.

Expansion

Expansion is a fluid dynamics process in which the condensed liquid refrigerant, under relatively high pressure from the compressor, is forced through an expansion valve into an area of substantially lower pressure. The expansion valve acts as a nozzle, constricting and then accelerating the liquid refrigerant until it passes through a threshold where the constriction is removed and rapid expansion occurs. At this point, the expanding refrigerant enters the evaporator coil, where it will cool and change state again.

Evaporation

Evaporation is a fluid dynamics process in which the rapidly expanding refrigerant liquid changes into a vapor. As the liquid enters the evaporator coil, it also enters an area of substantially lower pressure. As a result, it vaporizes, and in the process absorbs heat. Air in the controlled environment is circulated across this coil, which, in turn, absorbs heat from the air. The fan distributes the resulting cooler air throughout the controlled environment. The refrigerant vapor in the evaporator coil is then drawn back through the closed loop to the compressor, at which point the entire cycle repeats.

Conclusion

The operation of an air conditioner involves four steps. First, the compressor pumps refrigerant under pressure into the condenser coil. Here it is liquefied, giving up heat that is removed by a fan circulating air over the condenser coil. The liquid refrigerant moves through a closed loop, through an expansion valve, and into the lower-pressure evaporator coil. Here, the refrigerant changes into a vapor, absorbing heat from air passing over the evaporator coil. This air then cools the controlled environment, while the refrigerant is drawn back into the compressor. At this point the cycle is complete, and the process repeats.

Description of a Conceptual Process

The following is a brief, notional example of an iterative, conceptual process. Computer programmers use this particular process, which is roughly analogous to an insertion sort algorithm, to put lists of unsorted data into either ascending or descending order. In reality, the actual implementation of an insertion sort algorithm in modern, higher-level languages is far more complex and efficient than what is represented here.

Putting It All Together This example, which integrates two visuals into the description, is written for a general, technical audience that does not have specific expertise in computer programming and information systems. For illustration, this example process for sorting data in descending order uses the following list of six numbers:

> 6, 32, 8, 19, 3, 20

Process Description of an Insertion Sort

Introduction

Insertion sort is an iterative process in which a list of unsorted data is placed in either ascending or descending order. Insertion sort gets its name from the way in which the data are sorted. During each iteration, the smallest value from a list of unsorted numbers is identified and then inserted into the front of the list. As the process continues, the sorted area in the front of the list gets larger, while the remaining unsorted area gets smaller. At some point, when all the numbers from the unsorted list have been inserted, the list is sorted, and the process is complete. The insertion sort process involves the following iterative steps: evaluation, identification, and insertion. See Figure 4.2 for a flowchart of the process.

Discussion
Evaluation

Evaluation is the scanning process in which the list of numbers is tested for proper order. This

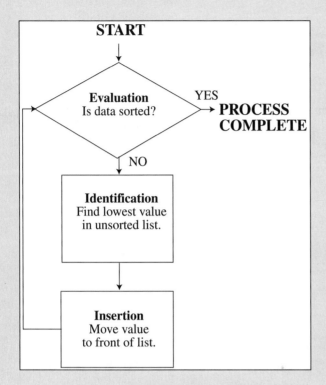

Figure 4.2
Sort process
flowchart

step normally involves a pass through the list where a series of comparisons of adjacent numbers determines whether the list is ordered. If the list is ordered, the process is complete and ends. If the list is not ordered, then the lowest value is identified.

Identification

Identification is the scanning process by which the lowest value in the list is located. Normally, the lowest value is identified by a series of comparisons and swaps. Once identified, this value can be inserted.

Insertion

Insertion is a relocation process in which the lowest value in the unsorted area of the list is placed

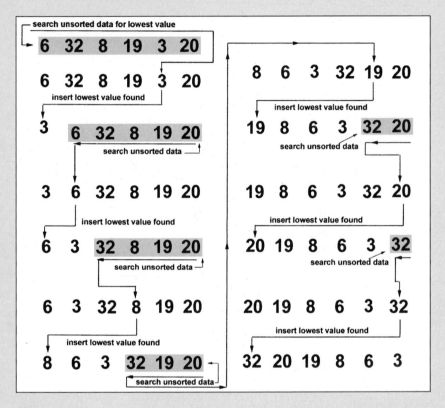

Figure 4.3
Insertion sort example

or *inserted* at the front of the list. On the first iteration, this value starts the ordered or sorted area of the list, while the remaining numbers constitute the unordered or unsorted area of the list. Once the lowest value has been inserted, the process returns to the evaluation step, where the iteration continues until no unordered numbers remain (see Figure 4.3 for an example).

Conclusion

Insertion sort is an iterative process in which a list of unsorted data is placed in either ascending or descending order. During each iteration, the list is evaluated to see if it is, in fact, ordered. If not, the smallest value from a list of unsorted numbers is identified and then moved to the front of the list. As the process continues, the sorted area in the front of the list gets larger, while the remaining unsorted area gets smaller. At some point, when all the numbers from the unsorted list have been moved, the list is sorted, and the process is complete. Insertion sort is one of the many algorithms used by computer programmers to sort lists of data.

Process Description Checklist

- Have I defined the mechanism or the process, and have I extended this definition with any theory or principles necessary for my reader's understanding? Have I included theory or principles that either are not needed or are above or below the level of my reader?
- Have I described the process's overall function or purpose?
- Have I described the overall function or operation involved in the process?
- Have I listed the main steps of the process in the introduction?
- Have I defined each of these steps?
- Have I described each step's function or purpose?
- Have I discussed the needed theory or operating principles for each step?
- Have I precisely described what happens in each step?
- Have I provided transitions from one step to the next step?
- For iterative processes, have I clearly shown when branching occurs and when the process is complete?
- Have I concluded by summarizing the process's purpose and function?
- Have I given a sense of finality to the description?

Exercise   Apply the checklist just given to the following process description of the QuadFINKEL figure skating jump. (Of course I named the jump after myself. If Ulrich Salchow, Alois Lutz, and Axel Paulson can do it, so can I—except, of course, I do not know how to skate.) In particular, look at the following:

- First, is the subject a process? If so, what areas of this process description are not technical writing? Do any areas deal with abstract, imprecise value judgments and concepts?
- For what audience was this description written? Is there a consistent audience? What areas of expertise would be required for the reader to understand the various parts? At what level of expertise is this paper written?
- How is the paper organized? Does the introduction properly introduce the process? Are the steps well defined? Are they adequately described? Does each step transition well to the next step?
- Look at the visuals. One exists for each step of the process. Are these visuals useful? Are they properly marked? Do they have adequate labeling? Are they placed effectively in the document? Are they well integrated into the text's description?

Process Description of a QuadFINKEL

Introduction

In figure skating, the *QuadFINKEL* is a competitive jump that combines a Lutz glide, Axel vault, high-speed spin, and Salchow landing. The QuadFINKEL is regarded by the skating world as the most difficult maneuver in the history of the sport. The QuadFINKEL, which requires a skater of great power, skill, and courage, has six steps: the glide, the turn, the vault, the spin, the stall, and the landing (see Figure 4.4).

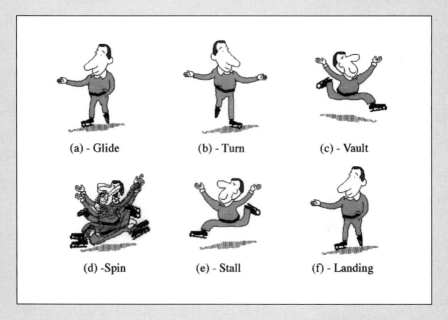

(a) - Glide   (b) - Turn   (c) - Vault

(d) -Spin   (e) - Stall   (f) - Landing

Figure 4.4
QuadFINKEL

Discussion

The Glide

The *glide* is an acceleration maneuver by which the skater establishes high-speed movement on the right outside back edge. As the glide speed approaches 55 mph, the left hand is inserted into the left pocket to reduce aerodynamic drag, while the right hand is extended to provide rudder control; see Figure 4.4(a). When backward velocity reaches 60 mph, the skater is ready to turn.

The Turn

The *turn* is a transition maneuver in which the skater's body rotates 180 degrees counterclockwise while maintaining the original vector. Additionally, the skater shifts from the right back outside edge to the left front outside edge and spreads both arms in preparation for the vault; see Figure 4.4(b).

The Vault

The *vault* is a propulsion maneuver in which the skater leaps off the right front outside edge, accomplishing an Axel-entry blade jump. Once airborne, the skater extends both hands up and away from the body to provide rotational stability and artistic elegance; see Figure 4.4(c). This action prepares the skater for the spin.

The Spin

The *spin* is a rotational maneuver in which the airborne skater rapidly moves his or her extended arms and hands to create both substantial artistic impression and a low-pressure vortex with resulting counterclockwise torque; see Figure 4.4(d). Generally, spin speeds exceeding 2,000 rpm are possible using this technique. The skater then turns his or her arms to simulate a Bernoulli surface, which provides sufficient aerodynamic lift to sustain autogyrational flight. (Note: In the Southern Hemisphere the rotation would be clockwise, so the QuadFINKEL is not possible south of the equator.) Once the rotational torque is established and the skater has made $3\frac{1}{2}$ revolutions, he or she is ready for the stall.

The Stall

The *stall* is an aerodynamic maneuver in which the skater uses his or her outstretched hands as spoilers to slow rotation to the point where lift is no longer sufficient to sustain flight; see Figure 4.4(e). This action is taken after $3\frac{1}{2}$ revolutions, so that a stall condition occurs at precisely $4\frac{1}{2}$ revolutions. The stall leads immediately to landing.

The Landing

Landing is the final jump maneuver by which the skater touches down on the right back outside edge; see Figure 4.4(f). The skater's left hand is reinserted in the pocket while the outstretched right hand is used again for rudder control. The left edge must not touch the ice until the skater's speed has been reduced to zero. At this point, the QuadFINKEL jump is complete.

Conclusion

The QuadFINKEL is the most difficult of all competitive figure skating jumps. To do a Quad-FINKEL, the skater does an Axel vault off the left front outside edge after a Lutz acceleration glide on the right outside back edge. The jump uses sophisticated aerodynamic forces to sustain a high-speed spin, followed by a Salchow landing on the right back outside edge. Because of its hypnotic artistic impression and demanding technical qualities, it has been said that "one who achieves the QuadFINKEL always lands on the gold."

5

Proposals

Proposals are among the most important documents one can write. Persons and organizations that write effective proposals win grants, contracts, and jobs; persons and organizations that do not write effective proposals often just wind up "going away"—sometimes "far away." Proposals are important because they, directly or indirectly, provide the income that keeps us warm, dry, and well fed!

What Is a Proposal?

Proposals are specialized, technical business documents that offer persuasive solutions to problems. All technical documents are designed to communicate ideas objectively and clearly. Your goal in writing a proposal, however, goes beyond just precise communication. A proposal also needs to *sell* the reader on some idea—usually that he or she (or his or her organization) needs specific goods or services that you (or your organization) can provide.

To be successful, you normally need to do at least three things in any proposal you write:

1. Describe, identify, or refer to a problem that needs to be solved. The reader probably already knows that he or she has a problem. You still need to describe it, however, because the reader may not fully understand or appreciate the scope, magnitude, or complexity of the issues at hand. Additionally, by providing this description, you establish credibility by showing your reader that you understand the problem.

2. Offer a viable solution for the problem. You have to demonstrate to the reader that your proposed approach will, in fact, solve the problem effectively and efficiently.

3. Show that you can effectively implement this solution. The fact that you have an effective solution does not mean much if you cannot implement it. That is why, in any proposal, you must show that you have the skills and resources required to do what you are proposing. One of the best ways of demonstrating your capability is with successful prior performance. In other words, if you are proposing to design a Web page for your company, it would be a definite plus if you could point to other successful Web pages that you have designed. Showing that you have done the kind of thing you are proposing to do—and have done it well—can be persuasive.

Formal and Informal Proposals

Proposals are generally categorized as *formal* or *informal*. Formal proposals are normally large, comprehensive documents produced by a team of experts on behalf of an organization. Informal proposals are generally short documents of limited scope written by an individual.

Formal Proposals

The topic of formal proposals is well beyond the scope of this book. But what if your boss just told you to put together a formal proposal, and you have no clue what to do? Well, quite honestly, your boss is a fool, and you are using the wrong book! Formal proposals are lengthy projects undertaken by skilled proposal writers functioning together in a well-structured, proposal team environment. These documents are prepared in response to a formal request for proposal (RFP). Proposal teams take great care to respond to each RFP requirement.

Formal proposals can take many forms, but a typical one might include the following:

- An "executive summary" that synopsizes the substance of the proposal. This section is usually written by a senior decision maker such as a company's vice president. This executive summary is the only part of the proposal that many readers look at, so it can be very important. Executive summaries are discussed in more detail in Chapter 18.
- A "technical volume" that lays out the proposed solution in detail. This section is normally written by a team of engineers and scientists who are responsible for solving the problem.
- A "management volume" that describes the organizational structure and key players who will implement the proposal if accepted. This section is generally written by a team of experts in management theory and organizational structure.
- A "cost volume" that provides detailed analysis and data regarding the cost of implementing the proposed solution. This section is often written by a team of financial planners, auditors, comptrollers, and accountants.
- A "resources volume" that provides detailed analysis and data regarding both the human and physical resources required to implement the proposed solution. This section is written by human resource experts, who hire the necessary skills, and by facilities experts, who locate, modify, or build the needed facilities. In some cases this section is included in the "management volume."

Formal proposals frequently take months to produce at a cost of tens of thousands of dollars.

Writing these proposals is a difficult task, especially because these documents are evaluated in an extremely competitive environment. Winning or losing is based on the quality of the proposed solution, the credibility of the proposing organization, and the estimated cost of implementing the proposal. In the end, the proposal representing the best overall value normally wins the contract.

Government Proposals

Perhaps the best example of just how challenging proposal writing can be involves writing proposals for the U.S. government, where federal tax dollars are involved. Writing proposals for the government is truly an exacting, unforgiving activity. The content, layout, and procedures are precisely spelled out, and evaluation of proposals is rigorous and comprehensive. Source selection panels composed of technical, management, human resource, financial, and other experts evaluate each proposal. They aggressively seek out all discrepancies between tasks and skills, skills and costs, management philosophy and organizational structure, and so on. They score each proposal based on the effectiveness of the solution, the risks involved, the costs in both time and dollars, and the capabilities and past performance of the proposing company. The goal of the source selection process is to select the proposal that offers the best value to the government.

Interaction between proposal writers and proposal evaluators (that is, between the company and the government) is strictly controlled. Even the appearance of impropriety can have serious legal implications. After preliminary interactions to evaluate the objectives of the project, the government issues an RFP that precisely spells out the problem that needs to be solved and the parameters and constraints on any proposed solution. In some cases, proposals are presented as technical briefings (Chapter 15), or they may be submitted electronically (Chapter 16). Larger projects may require multivolume documents.

In most cases, government proposals have specific deadlines and delivery requirements, and must precisely meet all format and content specifications.

Informal Proposals

The informal proposal is the primary focus of this chapter. Written by individuals, not teams, informal proposals typically address a limited problem for which a relatively straightforward solution exists. Frequently informal proposals take the form of a long letter or a short document.

Informal proposals may also be either solicited or unsolicited. A *solicited proposal* responds to a specific, often written, request. A homeowner might ask a contractor for a proposal to replace the roof on a house, or a company might ask a lighting firm for a proposal to illuminate a manufacturing area. With a solicited proposal, the problem has already been identified, and the decision to solve the problem has already been made.

Unsolicited proposals, on the other hand, are proposals that no one has asked for and, perhaps, that no one wants. The recipient has not decided to solve a problem and may not even realize that a problem exists—or may not want to realize it. Unsolicited proposals often come from within an organization—for example, an assembly line worker may send the supervisor a letter suggesting a change to a manufacturing process. Unsolicited proposals, as you might imagine, have less chance of being accepted.

Informal proposals can take many forms and can be organized in many different ways. The best advice is to carefully think through all aspects of the problem and your proposed solution, and to use your common sense. Also, follow any guidelines from your boss or an agency requesting the proposal.

If you do not have specific guidance, then Outline 5.1 provides a fairly typical approach you can use to organize informal proposals. We will look at each element of this outline and see how you might go about putting together such a proposal.

As a hypothetical example, imagine that the International Olympic Committee (IOC) has asked your organization, CompuLAB, to provide

Outline 5.1 Informal Proposals

Introduction
• Purpose	What is the purpose of this proposal?
• Background	What is the problem that needs to be solved?
• Scope	What are the limitations on this proposal and why?

Discussion
• Approach	What is the proposed solution?
• Result	How will the solution solve the problem?
• Statement of work	What tasks will be performed as part of this solution?

Resources
• Personnel	Who will be doing the work and why are they qualified?
• Facilities and equipment	What physical resources will be used to do the work?

Costs
• Fiscal	How much money will the proposed solution require?
• Time	How much time will the proposed solution require?

Conclusion
• Summary	What are the benefits and risks of adopting this proposal?
• Contact	Whom should one contact for more information?

computer modeling and analysis of the Quad-FINKEL figure skating jump (the Chapter 4 exercise). To respond to the IOC, your CompuLAB division must create an informal (that is, internal) proposal to upgrade its computer systems. The following proposal, written for the experts at CompuLAB, is geared to an expert audience knowledgeable in modeling and analysis, in the technology used to accomplish such modeling and analysis, and in the overall resources available to CompuLAB.

Introduction

Purpose. Tell the reader why you have written this document. The reader may not know or fully understand the purpose. So be specific, and if you are responding to a particular request, say so. For example, what if the purpose were described as follows?

> This document proposes a general system upgrade for CompuLAB's Sports Analysis Measurement Division (SPASM-D).

Pretty useless! This purpose description does not give the reader adequate specifics to put the proposal in its proper context. Everyone at CompuLAB is looking for a systems upgrade. In fact, everyone everywhere is looking for a systems upgrade. This proposal might wind up in a pile with everyone else's "wish list." To fix this problem, we need only add specifics:

> This document proposes a general system upgrade for CompuLAB's Sports Analysis Measurement Division's (SPASM-D) computing capabilities. This upgrade is necessary to enable CompuLAB's response to the International Olympic Committee's (IOC) tasking for modeling and analysis of the QuadFINKEL figure skating jump.

Background.   In this section, describe the problem that needs to be solved, adding any background necessary to clarify the requirement or put it in the proper context. Including specifics also demonstrates your understanding of the problem and adds to your credibility:

> The QuadFINKEL figure skating jump is so demanding that the IOC is considering the award of an extra-large gold medal to anyone successfully landing the jump in Olympic competition—and, of course, surviving. The IOC has tasked SPASM-D, under IOC Contract IOC-135549, to accomplish advanced 3-D modeling, simulation, and analysis of the QuadFINKEL. This modeling and analysis would give the IOC the scientific basis for justifying the special award. This analysis must be completed within six months of final project approval.
>
> Modeling this skating jump is a complex process due to the jump's chaotic nature and high-speed dynamics. To accomplish this analysis in the specified timeframe, SPASM-D, in the attached Technical Report TR-193345, has identified the need for a stand-alone, state-of-the-art graphics, modeling, and analysis capability within the laboratory area. The report identifies the requirement for standard, high-speed TCP/IP interfaces via the Internet to various IOC activities. These stand-alone capabilities must interface through the existing 100 Base-T Fast Ethernet backbone to other laboratory resources. TR-193345 also stipulates that the cost of the upgrade should not exceed $22,500.

Scope.   Clarify exactly what your proposal covers. Remember: an accepted proposal may be considered a binding contract that obligates both parties. Be careful to spell out any exclusions in the scope section:

> This proposal addresses only the system upgrade of the SPASM-D Analysis Laboratory in support of the

IOC tasking for modeling and analysis of the QuadFINKEL. This proposal does not include other graphics, modeling, and analysis tasks.

Discussion

Approach. Use this section to lay out precisely how you are proposing to solve the problem. Provide enough detail to clearly demonstrate that you have researched the problem, that you understand it, and that you have developed an effective solution.

SPASM-D proposes to the CompuLAB Technical Review Committee that the SPASM-D laboratory's capabilities be upgraded and augmented with a dedicated local area network (LAN) composed of two Titanium Graphics 990 Visual Workstations, one Titanium Graphics 990 Modeling Station, and one CompuLAB SuperHUB 1000-X.

Titanium Graphics workstations provide a recognized standard of excellence in performing advanced graphics, modeling, and analysis tasks. Off the shelf, these systems have the required processing power, storage capacity, and network/device interfaces to function seamlessly on the existing network, as well as the Internet. By connecting these workstations with 100 Base-T Fast Ethernet through a CompuLAB SuperHUB, then cascading this SuperHUB to the company's network backbone, these workstations easily can provide the needed stand-alone computational and storage capabilities. They will also meet all company and Internet connectivity requirements.

Result. Use this section to show what benefits will accrue from the proposed solution:

The Titanium Graphics 990 Modeling Station will provide the dedicated analytical capabilities required to thoroughly model and understand the various components of the QuadFINKEL. The two Titanium

Graphics 990 Visual Workstations will provide the 3-D graphics rendering required by the IOC tasking.

Statement of Work. Use this section to describe the major tasks you will perform to implement the proposed solution. In this example, one might identify three major tasks:

To achieve the goals of this proposal, the following tasks will be accomplished:
- Acquire the necessary equipment; transport, unpack, assemble, and place in the work area. (8 hours)
- Set up the operating systems and configure network connectivity. (16 hours)
- Install application software, check it out, and run calibration and verification simulations. (16 hours)

Resources

Personnel. In this section, discuss who will be doing the work and why they are qualified:

The SPASM-D simulation and modeling staff will analyze and model the QuadFINKEL using proprietary laboratory software. Additionally, CompuLAB graphics consultants and IOC figure skating experts will work closely with the simulation and modeling staff to assure accuracy and effectiveness of all required 3-D rendering.

Facilities and Equipment. Here you should describe the physical resources that will be used to do the work:

The SPASM-D facility includes adequate space for this effort. Suite 104 in Building 45 has the required network access and is available for this project.

The specific computer equipment required includes the following:

- Two Titanium Graphics 990 Visual Workstations*
- Integrated Visual Computing architecture with Argon graphics chipset
- Notel Octagon III Xeon 450 MHz processor with 512K L-2 cache
- 128 MB ECC SDRAM
- 9 GB Ultra2 SCSI drive
- Two 64-bit PCI buses
- 1.44 MB floppy drive, 32X CD-ROM drive
- Integrated 10/100 Fast Ethernet
- IEEE-1394, parallel, serial, USB, video, and audio ports
- Microsoft® Windows NT 4.0
- Three-year warranty with one-year on-site service

- One Titanium Graphics 990 Modeling Station
- Notel Octagon III Xeon 450 MHz processor with 512K L-2 cache
- 256 MB ECC SDRAM
- 9 GB Ultra2 SCSI drive
- Titanium Graphics 1600W 17.3-inch SuperWide digital flat panel monitor

- One CompuLAB SuperHUB 1000-X
- Cat 5 100 Base-T cable with connectors

*Note: existing CompuLAB display monitor resources are available to support these workstations.

Costs

Fiscal. Be sure your cost estimate falls within the monetary requirements and constraints of the problem. In this case, the background discussion specified a maximum cost of $22,500.00.

The proposed system upgrade includes the following equipment and installation costs:
Equipment:
- Two TGI 990 Visual Workstations
  @ $5,995.00 ea. = $11,990.00
- One TGI 990 Modeling Station
  @ $9,090.00 ea. = $ 9,090.00
- One SuperHUB X-1000
  @ $ 200.00 ea. = $ 200.00

- 400 feet Cat 5 100 Base-T cable with connectors
  @ $ 000.19 ft. = $ 76.00

Installation:

- .02 full-time equivalent (FTE) technician (40 hours)
  @ $42,000/FTE = $840.00

Total cost: $22,196.00

Time.   Be sure your time estimate falls within the requirements and constraints of the problem. In this case, the background discussion specified a maximum time of six months to complete the analysis. The schedule for the proposed upgrade has to meet that requirement.

Assuming availability of equipment and materials, and using fully qualified CompuLAB technicians, the entire upgrade can be completed in 40 hours. This estimate includes 8 hours to acquire, deliver, and set up the equipment; 16 hours to set up the operating system and network connectivity; and 16 hours to configure and check out the applications. This upgrade schedule would provide adequate time to accomplish the modeling and analysis required by the IOC tasking.

Conclusion

Summary.   Pay particular attention to this last section. It represents your final opportunity to sell the proposed solution by describing the benefits to be gained through your proposal. This section also allows you to demonstrate your competence by describing how your proposal addresses any risks inherent in the project.

The proposed Titanium Graphics LAN upgrade will provide a viable, cost-effective solution to meeting the IOC QuadFINKEL modeling and analysis requirement. The entire system can be up and running within a week for a price that is well within the cost guidelines. Given the quality of the system and reputation of its manufacturer for setting the standard for high-end analysis and modeling computing systems, the risks of this solution are minimal. In fact, the proposed system provides exactly the right capabilities at precisely the right time, with a cost and timeline that are well within the company's needs.

Contact. Tell the reader whom to contact for more information. Be sure this contact information is accurate and that the person specified understands the proposal and is available to answer questions.

For more information regarding this proposal, contact Edward R. Ronaldson, Ph.D., P.E., Director of Systems Engineering, SPASM-D, at Ext. 445; or e-mail eronald@CompuLAB.com.

Layout and Presentation

There is an unwritten rule that goes something like this: "Junk presented professionally often counts for more than junk presented unprofessionally." Many people present foolish ideas so smoothly that they convince others anyway. However, this "style-over-substance" approach is inappropriate and ineffective for technical writing.

You must present your substantive ideas as professionally as possible. Good ideas presented in an unprofessional manner are often interpreted as deficient—especially in technical proposals. To be persuasive, you must present valid ideas clearly and accurately in a document that is professional in every respect, including layout, style, and appearance.

Cover Letters and Title Pages

When submitting your proposal, you may want to include a cover letter (often called a transmittal letter). Cover letters ensure that the proposal gets to the right place and is considered in the proper context. These letters describe the enclosed document; the problem or requirement that the document addresses; the intended recipient; and the contact for any additional information. Here is a sample transmittal letter for the CompuLAB proposal:

CompuLAB
Sports Analysis and Measurement Division (SPASM-D)

March 31, 2000

Albert M. McEntyre, Chair
CompuLAB Technical Review Committee
1 Research Plaza, Suite 400
New York, NY 12091

Dear Dr. McEntyre:

Enclosed is the proposal for upgrading SPASM-D modeling and analysis capabilities in response to SPASM-D Technical Report TR-193345 requirements. This upgrade is required to accomplish advanced modeling and analysis of the QuadFINKEL figure skating jump in accordance with International Olympic Committee (IOC) tasking under Contract IOC-135549.

If you need additional information, please contact Edward R. Ronaldson, Ph.D., P.E., Director of Systems Engineering, SPASM-D, at Ext. 445; or e-mail eronald@CompuLAB.com.

Sincerely,

Evan J. Sonasky
Chief, Modeling and Analysis

EJS/mms
Enclosure

Additionally, when using a cover letter, you should also include a formal title page as the front page of your report. Title pages vary considerably in format, but they generally contain the following information:

- The complete title of the report.
- The name(s) of the author(s).
- The date of submission or period covered by the report.
- The name of the submitting organization.
- The name of the receiving organization.

Title pages may also need to contain additional information, such as RFP numbers and dates, as well as corporate trademarks and designs.

Here is a simple example of a title page:

Upgrading Modeling and Analysis Capability
A Proposal
by

CompuLAB SPASM-D/LXR

Submitted
on
March 31, 2000

for
The CompuLAB Technical Review Committee

Because proposals are persuasive documents designed to sell goods and services, you should not overlook the potential benefit of adding attachments to your report. Such attachments can provide information that supplements or clarifies material in the report but may be too detailed, extensive, or specialized to be included in the report itself. Attachments can be individually numbered and simply attached at the end of the report, or they can be included in a formal appendix section. If placed in an appendix,

Attachments and Appendixes

attachments should be identified individually as Appendix A, Appendix B, and so on.

Attachments can provide a wide range of information. For example, you might use attachments to show specific delivery schedules, detailed cost analysis, or additional background on you or your company. You could even attach additional information that supports your proposal—perhaps a detailed listing of successfully completed projects to establish prior performance. Think about what additional information you can include that will make your proposal more attractive to a potential client.

Putting It All Together

The following document provides a generic example of an internal, informal proposal. As such, it includes neither a letter of transmittal nor a title page; instead, this document uses a memorandum format for the header. This approach is common for *internal* documents. Remember, however, that *external* documents *always* require a cover letter and title page. When in doubt, it never hurts to include both. Also, this internal, informal proposal does not have any attachments.

Here is the entire proposal, then, assembled as it might be submitted.

> **DATE:** March 31, 2000
> **TO:** The CompuLAB Technical Review Committee
> **FROM:** CompuLAB SPASM-D/LXR
> **SUBJECT:** Proposal for Upgraded Modeling and Analysis Capability
> *Reference:* International Olympic Committee (IOC) Contract IOC-135549

1. Introduction

1.1 Purpose

This document proposes a general system upgrade for the CompuLAB's Sports Analysis Measurement Division's (SPASM-D) computing capabilities. This upgrade is necessary to enable Compu-LAB's response to the International Olympic Committee's tasking for modeling and analysis of the QuadFINKEL figure skating jump.

1.2 Background

The QuadFINKEL figure skating jump is so demanding that the IOC is considering the award of an extra-large gold medal to anyone successfully landing the jump in Olympic competition—and, of course, surviving. The IOC has tasked SPASM-D, under IOC Contract IOC-135549, to accomplish advanced 3-D modeling, simulation, and analysis of the QuadFINKEL. This modeling and analysis would give the IOC the scientific basis for justifying the special award. This analysis must be completed within six months of final project approval.

Modeling this skating jump is a complex process due to the jump's chaotic nature and high-speed dynamics. To accomplish this analysis in the specified timeframe, SPASM-D, in the attached Technical Report TR-193345, has identified the need for a stand-alone, state-of-the-art graphics, modeling, and analysis capability within the laboratory area. The report identifies the requirement for standard, high-speed TCP/IP interfaces via the Internet to various IOC activities. These stand-alone capabilities must interface through the existing 100 Base-T Fast Ethernet backbone to other laboratory resources. TR-193345 also stipulates that the cost of the upgrade should not exceed $22,500.

1.3 Scope

This proposal addresses only the system upgrade of the SPASM-D Analysis Laboratory in support of the IOC tasking for modeling and analysis of the QuadFINKEL. This proposal does not include other graphics, modeling, and analysis tasks.

2. Discussion

2.1 Approach

SPASM-D proposes to the CompuLAB Technical Review Committee that the SPASM-D laboratory's capabilities be upgraded and augmented with a dedicated local area network (LAN) composed of two Titanium Graphics 990 Visual Workstations, one Titanium Graphics 990 Modeling Station, and one CompuLAB SuperHUB 1000-X.

Titanium Graphics workstations provide a recognized standard of excellence in performing advanced graphics, modeling, and analysis tasks. Off the shelf, these systems have the required processing power, storage capacity, and network/device interfaces to function seamlessly on the existing network, as well as the Internet. By connecting these workstations with 100 Base-T Fast Ethernet through a CompuLAB SuperHUB, then cascading this SuperHUB to the company's network backbone, these workstations easily can provide the needed stand-alone computational and storage capabilities. They will also meet all company and Internet connectivity requirements.

2.2 Result

The Titanium Graphics 990 Modeling Station will provide the dedicated analytical capabilities required to thoroughly model and understand the various components of the QuadFINKEL. The two Titanium Graphics 990 Visual Workstations will provide the 3-D graphics rendering required by the IOC tasking.

2.3 Statement of Work

To achieve the goals of this proposal, the following tasks will be accomplished:
- Acquire the necessary equipment; transport, unpack, assemble, and place in the work area. (8 hours)

- Set up the operating systems and configure network connectivity. (16 hours)
- Install application software, check it out, and run calibration and verification simulations. (16 hours)

3. Resources

3.1 Personnel

The SPASM-D simulation and modeling staff will analyze and model the QuadFINKEL using proprietary laboratory software. Additionally, CompuLAB graphics consultants and IOC figure skating experts will work closely with the simulation and modeling staff to assure accuracy and effectiveness of all required 3-D renderings.

3.2 Facilities and Equipment

The SPASM-D facility includes adequate space for this effort. Suite 104 in Building 45 has the required network access and is available for this project.

The specific computer equipment required includes the following:

- Two Titanium Graphics 990 Visual Work-stations[*]
  - Integrated Visual Computing architecture with Argon graphics chipset
  - Notel Octagon III Xeon 450 MHz processor with 512K L-2 cache
  - 128 MB ECC SDRAM
  - 9 GB Ultra2 SCSI drive
  - Two 64-bit PCI buses
  - 1.44MB floppy drive, 32X CD-ROM drive
  - Integrated 10/100 Fast Ethernet
  - IEEE-1394, parallel, serial, USB, video, and audio ports
  - Microsoft® Windows NT 4.0
  - Three-year warranty with one-year on-site service

- One Titanium Graphics 990 Modeling Station
  - Notel Octagon III Xeon 450 MHz processor with 512K L-2 cache
  - 256 MB ECC SDRAM
  - 9 GB Ultra2 SCSI drive
  - Titanium Graphics 1600W 17.3-inch Super-Wide digital flat panel monitor
- One CompuLAB SuperHUB 1000-X
  - Cat 5 100 Base-T cable with connectors

*Note existing CompuLAB display monitor resources are available to support these workstations.

4. Costs

4.1 Fiscal

The proposed system upgrade includes the following equipment and installation costs:

Equipment:

- Two Titanium Graphics 990 Visual Workstations
          @ $5,995.00 ea.=$11,990.00
- One Titanium Graphics 990 Modeling Station
          @ $9,090.00 ea.= $9,090.00
- One CompuLAB SuperHUB X-1000
          @ $200.00 ea.= $ 200.00
- 400 feet Cat 5 100 Base-T cable w/connectors
          @ $ 000.19 ft. = $76.00

Installation:

- .02 full-time equivalent (FTE) technician
      (40 hours) @ $42,000/FTE = $ 840.00

Total cost: $22,196.00

4.2 Time

Assuming availability of equipment and materials, and using fully qualified CompuLAB technicians, the entire upgrade can be completed in 40 hours. This estimate includes 8 hours to acquire, deliver, and set up the equipment; 16 hours to set up the operating system and network connectivity; and 16 hours to configure and check out the applications. This upgrade schedule would provide adequate time to accomplish the modeling and analysis required by the IOC tasking.

5. Conclusion

5.1 Summary

The proposed Titanium Graphics LAN upgrade will provide a viable, cost-effective solution to meeting the IOC QuadFINKEL modeling and analysis requirement. The entire system can be up and running within a week for a price that is well within the cost guidelines. Given the quality of the system and reputation of its manufacturer for setting the standard for high-end analysis and modeling computing systems, the risks of this solution are minimal. In fact, the proposed system provides exactly the right capabilities at precisely the right time, with a cost and timeline that are well within the company's needs.

5.2 Contact

For more information regarding this proposal, contact Edward R. Ronaldson, Ph.D., P.E., Director of Systems Engineering, SPASM-D, at Ext. 445; or e-mail eronald@CompuLAB.com.

Proposal Checklist

- Have I defined the problem in enough detail to ensure that my reader will understand the context for this proposal?
- Have I described the background to this problem in enough detail to clearly identify the variables driving my proposed solution?
- Have I defined in the scope section how I am limiting my proposal?
- Have I laid out my proposed solution in adequate detail? Do I have enough detail to ensure that my solution is credible?

- Have I clearly described the benefits of my solution?
- Do I need to provide a statement of work? If so, have I adequately described the major tasks involved in implementing the proposed solution?
- Have I clearly defined the resources required to implement my proposed solution, including people, facilities, and equipment?
- Have I provided the required budget for implementing this proposal, including the costs in both money and time? Have I broken out the costs in adequate detail for my audience? Are these cost estimates consistent with the financial constraints of the problem?
- Is my time estimate consistent with the tasks in my statement of work?
- Have I clearly summarized the proposal and provided a strong, concluding argument for its adoption?
- Have I provided an available and knowledgeable contact at the end of the proposal?

6

Progress Reports

Once a proposal has been accepted and work begins on a project, you may be required by the terms of the contract to give the client periodic reports on the status of the project. That means answering straightforward questions such as these: Are you on schedule? Are you within budget? Are any risks evident? If so, how do you plan to control them? What remains to be done? What is your plan for doing it? What is your overall assessment?

Does the progress report requirement indicate that your client does not trust you? Maybe—but that is not the point. Normal business practice requires specific, written documentation, not abstract trust. In some cases, continuation of the contract or partial payments for work completed may depend on writing and submitting satisfactory progress reports.

On larger projects, progress or status reports may also be known as *milestone reports* because, like milestones on a road, they mark the passage from one point to another on a journey toward some final goal. Additionally, if your boss asks you for a report on a project that you are doing, often that type of progress report is called an *activity report*. In some cases you may be asked to write an activity report that documents the status of a range of projects you may be working on.

Progress reports, status reports, milestone reports, and activity reports all do basically the same thing and contain the same kinds of information. So to keep things simple, this chapter will focus only on progress reports.

| **What Is a Progress Report?** | Progress reports document the status of a project. These reports describe the various tasks that make up the project and analyze the progress that has been made toward completing each task. If you have written a proposal that has been accepted, then you have also probably committed yourself (or your organization) to completing certain tasks by a particular time and for a specified amount of money. Generally speaking, in a progress report you need to tell the reader three things: the problem you are solving, the solution you are implementing, and how well you are doing. You will find that writing a progress report is more pleasant when you have some progress to report. |

A progress report typically includes three parts; the first two parallel material in the original project proposal.

1. Review the problem that was the impetus for the original proposal. To do this, reference the original proposal by number or title, indicate when it was accepted, and then describe the problem that prompted the proposal in the first place. Often you can copy much of the problem description from the original proposal into the progress report.

2. Describe the solution offered in the original proposal, including the tasks involved in implementing this solution (usually listed in the statement of work) and the milestones (planned dates for completion) associated with each task.

3. Evaluate how well you are doing in terms of each task, and provide an overall assessment of

your progress. In other words, lay out to what extent you are getting the job done within the timeframe and cost constraints of the original proposal.

Progress reports can take many forms, from a simple letter to a multipage document. Use the format your boss tells you to use, or the one you are required to use by the client. If you do not have a specific format, Outline 6.1 should do the trick.

Progress Report Formats

To better understand how progress reports work, consider each part of this outline and how

Outline 6.1 Progress Reports

Introduction
- Purpose   Why are you writing this progress report?
- Background   What is the context for the project?
- Scope   What tasks of the project does this report cover?

Status
- Tasks completed
  - For each relevant task *(repeat this pattern for each task or activity completed)*
    - What was the task?
    - What was accomplished?
    - How long did it take?
    - What difficulties, if any, were encountered?
- Tasks remaining
  - For each relevant task *(repeat for each task or activity remaining)*
    - What is the task?
    - What has yet to be accomplished?
    - What are the timetable and strategy for doing it?
    - What are the risks and plans for mitigating them?

Conclusion
- Summary   What is your appraisal of the current status?
- Evaluation   How would you evaluate the progress made to date?
- Forecast   What is your forecast for completing this project?
- Contact   Who is the contact for this progress report?

you would use it. This example follows up Chapter 5's proposal for setting up the local area network to analyze and model the QuadFINKEL figure skating jump. Assume that the proposal was accepted and the project is under way.

The CompuLAB Technical Review Committee has asked for a short progress report that can be faxed to the International Olympic Committee for a meeting the day after tomorrow. No big deal! Producing this type of report is not difficult, especially if you have a copy of the original proposal handy. You will need that original proposal because it describes the context and tasks on which you will now report your organization's progress.

Introduction

Following Outline 6.1, first open the introduction by stating the purpose of this progress report. Think about the purpose carefully. Writing an effective report can be difficult if you cannot articulate why you are doing it.

> This report documents the progress CompuLAB has made on upgrading the modeling and analysis capability pursuant to its accepted SPASM-D/LXR proposal of March 31, 2000.

Next, in the background section of the introduction, describe the context for the project on which you are reporting. The best approach is to reference the accepted proposal, note its date of acceptance, and describe the specific problem it addressed, along with the proposed solution.

> The QuadFINKEL figure skating jump is so demanding that the IOC has tasked SPASM-D, under IOC Contract IOC-135549, to accomplish advanced 3-D modeling, simulation, and analysis of the QuadFINKEL. Modeling this skating jump is a complex process due to the jump's chaotic nature and high-speed dynamics.

To accomplish this analysis in the specified time-frame, SPASM-D proposed to the CompuLAB Technical Review Committee that the SPASM-D laboratory's capabilities be upgraded and augmented with a dedicated local area network (LAN). The new LAN will be composed of two Titanium Graphics 990 Visual Workstations, one Titanium Graphics 990 Modeling Station, and one CompuLAB SuperHUB 1000-X. The proposal was accepted without change on June 1, 2000, with the total system upgrade to be completed on or before June 30, 2000.

In the scope section of the introduction, describe the tasks or aspects of the project that this progress report covers. If the original proposal included a statement of work, you will probably report on some or all of the tasks listed. Specify what tasks you will be reporting on in this progress report. If applicable, specify which tasks you have reported on in an earlier report, which tasks you will report on in a subsequent report, and whether any tasks are recurring. If the original proposal did not include a statement of work, you will need to provide a more general description of the tasks on which you are reporting.

This report provides the status of all tasks described in the accepted proposal's statement of work, which includes the following:
- Acquiring the necessary equipment; transporting, unpacking, assembling, and placing in the work area. (8 hours)
- Setting up the operating systems and configuring network connectivity. (16 hours)
- Installing application software, checking it out, and running calibration and verification simulations. (16 hours)

Status

This section presents the status of each task listed in the scope section. Normally, you would treat completed tasks separately from the remaining tasks. The tasks completed section includes tasks that have been concluded and closed out. The tasks remaining section includes tasks that are still in progress and tasks that have not yet been started. Write the tasks completed section first, discussing each task separately:

Tasks Completed
Acquiring the necessary equipment; transporting, unpacking, assembling, and placing in the work area. (8 hours)

- All necessary equipment—including two Titanium Graphics 990 Visual Workstations, one Titanium Graphics 990 Modeling Station, and one Super-HUB 1000-X—was purchased on June 2, 2000, within budget estimates, and delivered, unpacked, and placed in Suite 104, Building 45, on June 4, 2000. The entire task required about seven hours. No difficulties were encountered.

Setting up the operating systems and configuring network connectivity. (16 hours)

- All three workstations came with Windows NT preinstalled. We successfully configured these machines and the hub to operate on the LAN. Initially the LAN experienced an excessive number of packet collisions. The problem was isolated to the SuperHUB, which was subsequently replaced. We completed system setup and network configuration in eight hours.

Once you've finished the tasks completed section, write the tasks remaining section in the same way:

Tasks Remaining
Installing application software, checking it out, and running calibration and verification simulations. (16 hours)

- We have just started this task. We are well ahead of schedule on implementing the statement of work, and we foresee no problems in completing this task and the entire project on time and within specified budget constraints.

Conclusion

The conclusion of the progress report summarizes and appraises of the progress to date and provides a forecast for the rest of the project (including any risks and plans for their mitigation). It also provides a contact in case more information is required. As in the proposal, the contact should be someone who is familiar with the project.

> The SPASM-D technical staff is very pleased with the progress made to date. We procured, transported, unpacked, assembled, and installed the equipment in its intended operating location in only seven hours, thereby completing Task 1 ahead of schedule. We configured the operating systems and achieved full network connectivity in 15 hours, thereby completing Task 2 on schedule. We are now in the process of installing, checking out, and calibrating the application software and expect to complete Task 3 within the scheduled 16 hours. Consequently, we at CompuLAB/SPASM-D have assessed the progress on this project as *excellent* and are confident that the entire project will be completed on time and within the proposed budget.
>
> For more information on this project, contact Edward R. Ronaldson, Ph.D., P.E., Director of Systems Engineering, SPASM-D, at Ext. 445; or e-mail eronald@CompuLAB.com.

Here is the entire progress report assembled as it might be submitted. Notice the addition of Figure 6.1, a timeline chart showing the relative progress for the three tasks. For short, informal reports, a timeline chart normally is not required; one is included here only as an example. Also notice that a memorandum header is used without a title

Putting It All Together

page or cover letter. As was the case with the original proposal (Chapter 5), this progress report is an informal, internal document. If it were designed for use outside the company, it would need a cover letter and title page.

DATE: June 7, 2000
TO: The CompuLAB Technical Review Committee
FROM: CompuLAB SPASM-D/LXR
SUBJECT: Progress Report for Upgraded Modeling and Analysis Capability
Reference: CompuLAB Proposal #CL-3478 of March 31, 2000

1. Introduction
1.1 Purpose
This report documents the progress CompuLAB has made on upgrading the modeling and analysis capability pursuant to its accepted SPASM-D/LXR proposal of March 31, 2000.

1.2 Background
The QuadFINKEL figure skating jump is so demanding that the IOC has tasked SPASM-D, under IOC Contract IOC-135549, to accomplish advanced 3-D modeling, simulation, and analysis of the QuadFINKEL. Modeling this skating jump is a complex process due to the jump's chaotic nature and high-speed dynamics. To accomplish this analysis in the specified timeframe, SPASM-D proposed to the CompuLAB Technical Review Committee that the SPASM-D laboratory's capabilities be upgraded and augmented with a dedicated local area network (LAN). The new LAN will be composed of two Titanium Graphics 990 Visual Workstations, one Titanium Graphics 990 Modeling Station, and one CompuLAB SuperHUB 1000-X. The proposal was accepted without change on

June 1, 2000, with the total system upgrade to be completed on or before June 30, 2000.

1.3 Scope

This report provides the status of all tasks described in the accepted proposal's statement of work, which includes the following:

- Task 1: Acquiring the necessary equipment; transporting, unpacking, assembling, and placing in the work area. (8 hours)
- Task 2: Setting up the operating systems and configuring network connectivity. (16 hours)
- Task 3: Installing application software, checking it out, and running calibration and verification simulations. (16 hours)

2. Status

2.1 Tasks Completed

- *Task 1:* Acquiring the necessary equipment; transporting, unpacking, assembling, and placing in the work area. (8 hours)
  - All necessary equipment—including two Titanium Graphics 990 Visual Workstations, one Titanium Graphics 990 Modeling Station, and one SuperHUB 1000-X—was purchased on June 2, 2000, within budget estimates, and delivered, unpacked, and placed in Suite 104, Building 45, on June 4, 2000. The entire task required about seven hours. No difficulties were encountered.
- *Task 2:* Setting up the operating systems and configuring network connectivity. (16 hours)
  - All three workstations came with Windows NT preinstalled. We successfully configured these machines and the hub to operate on the LAN. Initially the LAN experienced an excessive number of packet collisions. The problem was isolated to the SuperHUB, which was subsequently replaced. We completed

system setup and network configuration in eight hours.

2.2 Tasks Remaining

- *Task 3:* Installing application software, checking it out, and running calibration and verification simulations. (16 hours)
  - We have just started this task. We are well ahead of schedule on implementing the statement of work, and we foresee no problems in completing this task and the entire project on time and within specified budget constraints.

3. Conclusion

The SPASM-D technical staff is very pleased with the progress made to date (see Figure 6.1 for the project timeline). We procured, transported, unpacked, assembled, and installed the equipment in its intended operating location in only seven hours, thereby completing Task 1 ahead of schedule. We configured the operating systems and achieved full network connectivity in 15 hours, thereby completing Task 2 on schedule. We are now in the process of installing, checking out, and calibrating the application software and expect to complete Task 3 within the scheduled 16 hours. Consequently, we at CompuLAB/SPASM-D have assessed the progress on this project as *excel-*

Figure 6.1
Project timeline

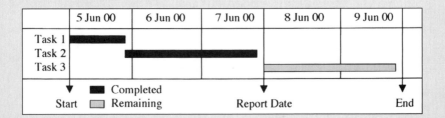

	5 Jun 00	6 Jun 00	7 Jun 00	8 Jun 00	9 Jun 00	
Task 1						
Task 2						
Task 3						

Start   ■ Completed   ☐ Remaining   Report Date   End

lent and are confident that the entire project will be completed on time and within the proposed budget.

For more information on this project, contact Edward R. Ronaldson, Ph.D., P.E., Director of Systems Engineering, SPASM-D, at Ext. 445; or e-mail eronald@CompuLAB.com.

- Have I specified the purpose, background, and scope of this report?
- Have I referenced the accepted proposal by name, number, and/or date?
- Have I reviewed the problem contained in that proposal?
- Have I reviewed the proposed solution to that problem?
- Have I specified the tasks that will be included in this report?
- Have I properly discussed the tasks completed and tasks remaining?
- Have I provided an appraisal and forecast in the conclusion?

Progress Report Checklist

7
Feasibility Reports

Feasibility reports are objective documents that recommend solutions to problems. In technical writing, feasibility reports address subjects that have well-defined parameters, including a problem, or multiple problems, that can be precisely described; and solutions, or multiple solutions, that can be objectively and empirically tested.

Feasibility reports are unbiased evaluations, although their conclusions and recommendations can, and frequently are, used to promote ideas and sell goods and services. Ideally, however, only someone who is totally impartial should write these reports. The author should have no stake in the outcome and should not care whether any or all of the solutions are adopted.

What Is a Feasibility Report?

Feasibility reports either determine the feasibility of solving a problem in a particular way, or recommend which of several options for solving a problem is the best approach. In the latter case, feasibility reports are often called recommendation reports. However, both feasibility and recommendation reports basically do the same thing: they objectively evaluate solutions.

Outline 7.1 provides a typical approach for organizing feasibility reports.

Outline 7.1   Feasibility Report

Introduction
- Purpose   What is the purpose of this feasibility report?
- Problem   What is the problem that needs to be solved?
- Scope   What are the alternatives and criteria?

Discussion
- Criterion #1
  - Explanation   What is the criterion, why was it selected, and how will it be used?
  - Data   What are the findings (data) for this criterion?
  - Interpretation   What is your interpretation of the data for this criterion relative to each alternative solution?
- Criteria #2–#*n*
  (Treat each remaining criterion in the same manner as Criterion #1.)

Conclusion
- Summary   What are the data and interpretations?
- Conclusions   What are your conclusions based on the data and interpretations?
- Recommendation   What is your recommended solution based on these conclusions?
- Contact   Who is the contact for this report?

Documentation   What sources and references were used?

Appendix   What materials are needed for support, but *not* for understanding the report?

Writing Feasibility Reports   Feasibility reports are organized in a straightforward and logical way. As Outline 7.1 shows, you need to do several things when writing a feasibility report:

1. *Define a problem that needs to be solved.* The difficulty here is that we are often solution-oriented; in many cases we skip the problem and go directly to the solution. What if a friend came to you and said, "I have a problem: I need to buy a computer, but I do not know which one to purchase"? The main problem your friend has

is that he or she does not have a problem. "I need to buy a computer" is a *solution* statement, not a *problem* statement. Why does your friend need a computer? *That* is the problem! To come up with candidate solutions, you would have to know why your friend needed a computer. Feasibility reports work exactly the same way. You cannot recommend a solution to a problem that is not clearly defined.

2. *Identify one or more candidate solutions.* This process can be tough; sometimes many more solutions exist than you will have the time or capability to evaluate. If you need a computer to surf the Net, how many choices do you have? This is almost like asking how many stars are in the Milky Way Galaxy. Coming up with just a few viable solutions can be challenging. Normally you can apply additional requirements to the existing problem that will allow you to narrow the list. Maybe you will buy only from an approved, local vendor; or you will shop only within five square miles of your home, or you will consider using only a certain catalog that gives you a rebate.

3. *Develop a set of criteria by which to objectively evaluate the candidate solutions.* The key here is "objective." Find meaningful measures that relate to the problem you have defined, and come up with a valid method for applying them. For example, when looking for a computer to surf the Net, you might use criteria such as cost, processor speed, monitor size and quality, reliability and warranty, bundled software, and included peripherals. These can be objectively described and measured. How attractive the case is would not be a good criterion because computer case attractiveness cannot easily be objectively described and measured.

4. *Collect application data for each criterion as it relates to each candidate solution, and then*

interpret that data. One thing that must be decided is how to weight the importance of each criterion. Sometimes the criteria can be weighted equally, but in many studies, some criteria are more important and need to count more in the final decision. For example, what if you wanted a computer to surf the Net using a satellite link from the back of an all-terrain vehicle in the upper Amazon region? In this case, reliability and maintainability might be far more important than, say, processor speed. On the other hand, if you planned to use the computer to do serious number crunching in the office, processor speed would be more important.

5. *Draw conclusions and make recommendations regarding the feasibility of the candidate solutions based on your interpretations.* The primary requirement here is objectivity and clear thinking. Look at the interpretations you have made for each criterion, and consider the relative weighting of the criteria. Always base your conclusions on this information—never on other information or considerations that are not fully treated in the report.

Here is a sample feasibility report that is based on Outline 7.1.

Introduction
Purpose

Describe the purpose of this report. This part is straightforward. What do you hope to achieve with this document? To what requirement does this document respond? Your purpose statement might look something like this:

> The purpose of this report is to recommend the best data backup system to support the Federated Scientific Associates (FSA) Web server. This report

responds to the FSA Senior Vice President for Information Systems' memo of July 15, 2000 (Appendix 1).

We know from this purpose statement that the report will focus on responding to the requirement for data backup of FSA's Web server. We also know that the original tasking memo for the report is available in Appendix 1. It is always a good idea to include the original tasking that requires (and justifies) a feasibility report containing recommendations for and against various options.

Once you have stated the purpose, you can define the problem addressed in this report.

Problem

You cannot write a feasibility report unless you have a problem. Solutions without problems are not very useful, and neither are solutions that are not geared specifically to solving problems. So to define the problem in this section, provide needed background and describe specific requirements that any viable solution must meet.

Consider the following problem statement:

FSA needs a backup system for its Web server.

This problem statement is useless because it does not actually state a problem; it puts forth a solution. That FSA needs a backup system is a solution to some problem that requires backups. The problem that requires backups should be the focus here, not the solution of providing backups.

A better problem statement might look something like the following:

The FSA Web server supports the FSA home page. The page, used primarily to provide on-line customer support and technical information, includes a dynamic technical database, a current list of customers with active support agreements, and a public download area for FSA software device drivers and updates. Recently, the Web page was expanded to include secure, on-line ordering of upgrades and peripherals.

Currently, FSA is using an Apple Macintosh, high-end G-3 Server with a 10 GB hard drive tied to the Internet through the company's dedicated T-3 connection. The site currently experiences about 10,000 hits, 3,000 page impressions, and 1,800 user sessions per week. With the addition of on-line ordering and the company's expanded use of its dynamic technical database, corporate headquarters has directed that the Information Systems Division (ISD) accomplish daily incremental backups, with monthly global backups of the entire site. Current estimates include the need to back up as much as 100 MB of data daily and up to 8 GB of data monthly.

Proprietary company server software currently provides the capability for automated, synchronized, mirrored, incremental backups to an external SCSI or IEEE 1394 (Firewire) device. Global backups must be accomplished manually. The selection of the external backup device is the focus of this report. The requirement is to provide reliable, robust backup capabilities in an economical way, with minimal loading of the Web server during backup operations.

Scope

The scope section lists the candidate solutions to be considered, along with the criteria that will be used to test each solution. First ensure that any candidate solution you consider is at least apparently viable. Some solutions may look good initially; however, when closely examined in terms of the problem that needs to be solved, these same ideas may appear silly. Take a preliminary look at each candidate solution to be sure it makes sense.

Also list the criteria by which you will evaluate each candidate solution. These criteria, driven by the problem, can include almost anything from cost, to reliability, to ease of use. Normally, in a technical paper, you would use only criteria that can be objectively and empirically quantified and tested. In the case of our example problem, we know that the solution must be robust, reliable, economically feasible, and efficient. Based on these requirements, we might develop a scope section that looks like the following:

Given that the backup device must be SCSI or Firewire compatible and that large amounts of data will be backed up routinely, this report will consider two general approaches: (1) an array of external, fixed-media hard drives; or (2) an external, removable-media hard drive. Optical rewritable drives are not considered because of initial cost and performance limitations.

The external, fixed- or removable-media hard drives should contain storage capacity for a minimum of six months of global backups, which could consume 48 GB of space, and two months of incremental backups, which could consume an additional 6 GB of space. Consequently, any solution must be capable of storing 54 GB of data at any given time.

Using approved FSA vendors, two options exist that will meet the requirements of the problem:

- Fixed media option (FMO): Purchase three DataStuff external, 18 GB, SCSI hard drives and chain them to the server's SCSI port.
- Removable media option (RMO): Purchase a single DataStuff 2 GB Removable External SCSI Drive, along with 24 2 GB cartridges for global backups and 3 2 GB cartridges for incremental backups.

This report will evaluate these two options using the following criteria: cost, performance, and reliability.

Discussion

The discussion section is the heart of the feasibility report. In this section you discuss the criteria, apply them to the candidate solutions, generate data, and interpret the results. *Notice that this section is organized by criteria, not solutions.* That is very important. You want to compare solutions for each criterion, not vice versa. If instead you organize the discussion around solutions, you will have to go through all of your criteria for each solution. By the time you get through all of the solutions, you will find it difficult to conceptualize and compare the solutions effectively because you will not remember which data pertained to which solution.

To begin the discussion section, here is the kind of treatment you might give the first criterion, cost.

Cost

First, in the explanation section, define the criterion, explain why you selected it (if necessary) and note any special weighting it will have in the evaluation process. If it has special weighting, explain how you chose your weighting system.

> *Explanation:* The *cost* of each item is the price that was published in FSA's *Coordinated Approved Vendor Catalog,* Number 42, Issue 14, June 2000. Company policy requires that the purchase be made from this catalog. Along with the other criteria, this criterion will be weighted equally as one-third of the total value. Although cost is considered important, the relatively small investments involved do not require additional weighting.

In the data section, discuss the findings using this criterion for each alternative solution. The superscript source reference after "Data" will be addressed in the documentation section of the feasibility report.

Data:[1]
FMO
Global and incremental BU requirement:
3 units @$729.00 each = $2,187.00
- Required cables and adapters = $50.00
- Total cost: $2,237.00
RMO
- Basic Unit: $349.00
- Disk requirement for global backups:
  24 disk cartridges @ $124.99 = $2,999.76
- Disk requirement for incremental backups:
  3 disk cartridges @ $124.99 = $374.97
- Required cables and adapters = $25.00
- Total cost: $3,748.73

In the interpretation section, discuss your interpretation of the data for this criterion for each alternative solution.

Interpretation: The FMO has clear cost advantages. Because of the high cost of removable media cartridges and the large numbers required by their relatively low storage capacity, implementing the RMO would cost about 67 percent more than implementing the FMO.

Next, in a similar manner, deal with the second criterion, reliability.

Reliability

Describe the criterion, explain why you selected it, and note any special weighting it will have in the evaluation process.

Explanation: Given the importance of providing backups for the Web server, the backup capability must be reliable. FSA must be able to depend on this backup capability to perform its function over an extended time without failure. To measure relia-bility, this study will consider the manufacturer's published mean time between failure (MTBF) sta-tistics and warranty for both the FMO and RMO. As with cost, this criterion is weighted at one-third the total value.

Provide the data and findings.

Data:[2]
FMO
- MTBF = 150,000 hours
- Warranty = 5 years on unit
RMO
- MTBF = 100,000 hours
- Warranty = 3 years on unit and lifetime on car-tridges

Discuss your interpretation of the data for this criterion for each alternative solution.

Interpretation: Both solutions are very reliable, although the FMO is more reliable. Assuming 8,760 hours in a year, and running full-time, the FMO would average a failure once every 17 years; the RMO would be slightly less reliable with a failure every 11 years. The FMO base unit also has the bet-ter warranty of five years, compared with the RMO's three-year warranty on the base unit.

The removable media are warranted for life; additionally, they provide inherent reliability by their sheer numbers. Unlike the FMO, the failure or loss of a single removable disk would have little impact on the backup archive. However, given the high MTBF for the FMO and the fact that the server itself provides redundancy to the FMO backup, statistically, the risk of using the FMO is very small.

Performance
Handle this criterion in a manner similar to the first two.

Explanation: For this study, *performance* refers to the speed at which the backup drive can read and write data. Speed can be expressed in both the access times for the disks and the data transfer rates. The better the performance, the lower the server loading during backup operations. However, because global backup operations will generally occur at low-usage hours, this criterion will not receive any additional weighting and will be given one-third of the total value.

Data:[3]
FMO
- Access = 7.5 ms seek time
- Transfer rate = 80 Mbps
RMO
- Access: 10 ms seek time
- Transfer rate = 8.7 Mbps

Interpretation: Both solutions would perform adequately in the backup operational profile. Although the relative access times differ somewhat, the FMO is substantially better in relative transfer rates, where the difference is almost an order of magnitude. This difference could be significant in terms of backup system performance and server load.

Conclusion

In the conclusion section, summarize the data and interpretations for all criteria and solutions. Do not include new information in the conclusion section; it should deal only with information already presented in the paper.

Summary

Briefly summarize the report and its findings.

This study evaluated both the FMO and the RMO based on three criteria: cost, reliability, and performance.

In terms of cost, the RMO costs 70 percent more than the FMO, although either option is affordable.

In terms of reliability, the RMO had a lower MTBF and shorter warranty, but the inherent redundancy of the relatively large number of removable media would tend to mitigate its lower MTBF. This mitigation is especially true because even the lower MTBF of 100,000 hours would still represent more than 11 years of constant operation.

In terms of performance, however, the FMO is clearly superior, with more than 10 times the data transfer rate. This result is significant.

Conclusions

Draw what overall conclusions you can from the material presented in the summary.

Both options are capable of providing cost-effective and reliable backup capabilities for the FSA Web server with adequate performance. However, the significant performance advantage of the FMO and its lower cost clearly make it the better choice.

Recommendation

The final step is to recommend the best option based on your conclusions.

The FMO, using three DataStuff external, 18 GB SCSI hard drives, is recommended for backing up the FSA Web server.

Contact

Provide the name of and contact information for someone who is knowledgeable in this technology, who is familiar with this feasibility report, and who is available to take questions and provide clarifications.

For additional information on this feasibility report, please contact Sheila A. Suppinsky, Senior Systems Administrator, Federated Scientific Associates, at (513) 873-5124, fax (513) 873-5009, or e-mail suppin@fsa_1.com.

Documentation

List the sources used in this report (see Chapter 13).

[1] Federated Scientific Associates, *Coordinated Approved Vendor Catalog,* No. 42, Issue 14, June 2000.

[2] DataStuff International, Technical Specifications, www.datastuff.com/tech/31203.html, updated November 28, 2000.

[3] DataStuff International, Technical Specifications, www.datastuff.com/tech/33298.html, updated November 28, 2000.

Appendix

Attach a copy of the original tasking letter referenced in the introduction.

Here is the entire feasibility report assembled as it might be submitted. Notice the addition of Table 7.1, a compilation of the data for each criterion as applied to each option; and Appendix 1, the original tasking document.

Putting It All Together

DATE: August 24, 2000
TO: FSA/HQ
FROM: FSA/ISD
SUBJECT: Feasibility Report for the FSA Web Server's Backup Capability
Reference: FSA/HQ Memo of July 15, 2000

1. Introduction

1.1 Purpose

The purpose of this report is to recommend the best data backup system to support the Federated Scientific Associates (FSA) Web server. This report responds to the FSA Senior Vice President for Information Systems' letter of July 15, 2000 (Appendix 1).

1.2 Problem

The FSA Web server supports the FSA home page. The page, used primarily to provide on-line customer support and technical information, includes a dynamic technical database, a current list of customers with active support agreements, and a public download area for FSA software device drivers and updates. Recently, the Web page was expanded to include secure, on-line ordering of upgrades and peripherals.

Currently, FSA is using an Apple Macintosh, high-end G-3 Server with a 10 GB hard drive tied to the Internet through the company's dedicated T-3 connection. The site currently experiences about 10,000 hits, 3,000 page impressions, and 1,800 user sessions per week. With the addition of on-line ordering and the company's expanded use of its dynamic technical database, corporate headquarters has directed that the Information Systems Division (ISD) accomplish daily incremental backups, with monthly global backups of the entire site. Current estimates include the need to back up as much as 100 MB of data daily and up to 8 GB of data monthly.

Proprietary company server software currently provides the capability for automated, synchronized, mirrored, incremental backups to an external SCSI or IEEE 1394 (Firewire) device. Global backups must be accomplished manually. The selection of the external backup device is the focus of this report. The requirement is to provide reliable, robust backup capabilities in an economical way, with minimal loading of the Web server during backup operations.

1.3 Scope

Given that the backup device must be SCSI or Firewire compatible and that large amounts of data will be backed up routinely, this report will consider two general approaches: (1) an array of external, fixed-media hard drives; or (2) an external, removable-media hard drive. Optical rewritable drives are not considered because of initial cost and performance limitations.

The external, fixed- or removable-media hard drives should contain storage capacity for a minimum of six months of global backups, which could consume 48 GB of space, and two months of incremental backups, which could consume an additional 6 GB of space. Consequently, any solution must be capable of storing 54 GB of data at any given time.

Using approved FSA vendors, two options exist that will meet the requirements of the problem:
- Fixed media option (FMO): Purchase three DataStuff external, 18 GB, SCSI hard drives and chain them to the server's SCSI port.
- Removable media option (RMO): Purchase a single DataStuff 2 GB Removable External SCSI Drive, along with 24 2 GB cartridges for global backups and 3 2 GB cartridges for incremental backups.

This report will evaluate these two options using the following criteria: cost, performance, and reliability.

2. Discussion
2.1 Cost

2.1.1 Explanation.   The *cost* of each item is the price that was published in FSA's *Coordinated Approved Vendor Catalog,* Number 42, Issue 14, June 2000. Company policy requires that the purchase be made from this catalog. Along with the other criteria, this criterion will be weighted equally as one-third of the total value. Although cost is considered important, the relatively small investments involved do not require additional weighting.

2.1.2 Data:[1]
FMO
- Global and incremental BU requirement:
  - 3 units @ $729.00 each = $2,187.00
- Required cables and adapters = $50.00
- Total cost: $2,237.00

RMO
- Basic unit: $349.00
- Disk requirement for global backups:
  - 24 disk cartridges @ $124.99 = $2,999.76
- Disk requirement for incremental backups:
  - 3 disk cartridges @ $124.99 = $374.97
- Required cables and adapters = $25.00
- Total cost: $3,748.73

2.1.3 Interpretation.   The FMO has clear cost advantages. Because of the high cost of removable media cartridges and the large numbers required by their relatively low storage capacity, implementing the RMO would cost about 67 percent more than implementing the FMO.

2.2 Reliability

2.2.1 Explanation.   Given the importance of providing backups for the Web server, the backup capability must be reliable. FSA must be able to depend on this backup capability to perform its function over an extended time without failure. To measure reliability, this study will consider the manufacturer's published mean time between failure (MTBF) statistics and warranty for both the FMO and RMO. As with cost, this criterion is weighted at one-third the total value.

2.2.2 Data:[2]
FMO
- MTBF = 150,000 hours
- Warranty = 5 years on unit

RMO

- MTBF = 100,000 hours
- Warranty = 3 years on unit and lifetime on cartridges

2.2.3 Interpretation. Both solutions are very reliable, although the FMO is more reliable. Assuming 8,760 hours in a year, and running full-time, the FMO would average a failure once every 17 years; the RMO would be slightly less reliable with a failure every 11 years. The FMO base unit also has the better warranty of five years, compared with the RMO's three-year warranty on the base unit.

The removable media are warranted for life; additionally, they provide inherent reliability by their sheer numbers. Unlike the FMO, the failure or loss of a single removable disk would have little impact on the backup archive. However, given the high MTBF for the FMO and the fact that the server itself provides a redundancy to the FMO backup, statistically, the risk of using the FMO is very small.

2.3 Performance

2.3.1 Explanation. For this study, *performance* refers to the speed at which the backup drive can read and write data. Speed can be expressed in both the access times for the disks and the data transfer rates. The better the performance, the lower the server loading during backup operations. However, because global backup operations will generally occur at low-usage hours, this criterion will not receive any additional weighting and will be given one-third of the total value.

2.3.2 Data:[3]
FMO
- Access = 7.5 ms seek time
- Transfer Rate = 80 Mbps

RMO
- Access: 10 ms seek time
- Transfer rate = 8.7 Mbps

2.3.3 Interpretation. Both solutions would perform adequately in the backup operational profile. Although the relative access times differ somewhat, the FMO is substantially better in relative transfer rates, where the difference is almost an order of magnitude. This difference could be significant in terms of backup system performance and server load.

3 Conclusion

3.1 Summary

This study evaluated both the FMO and the RMO based on three criteria: cost, reliability, and performance. Table 7.1 provides a summary of these comparisons.

In terms of cost, the RMO costs 67 percent more than the FMO, although either option is affordable.

In terms of reliability, the RMO had a lower MTBF and shorter warranty, but the inherent redundancy of the relatively large number of

Table 7.1 Summary

	Fixed media option (FMO)	Removable media option (RMO)
Cost (U.S. dollars)	$2,237.00	$3,748.73
Reliability		
MTBF (hours)	150,000 hours	100,000 hours
Warranty term (years)	5 years	3 years (lifetime on media)
Performance		
Access (seek time in ms)	7.5 ms	10 ms
Transfer rate (Mbps)	80 Mbps	8.7 Mbps

removable media would tend to mitigate its lower MTBF. This mitigation is especially true because even the lower MTBF of 100,000 hours would still represent more than 11 years of constant operation.

In terms of performance, however, the FMO is clearly superior, with more than 10 times the data transfer rate. This result is significant.

3.2 Conclusions

Both options are capable of providing cost-effective and reliable backup capabilities for the FSA Web server with adequate performance. However, the significant performance advantage of the FMO and its lower cost clearly make it the better choice.

3.3 Recommendation

The FMO, using three DataStuff external, 18 GB SCSI hard drives, is recommended for backing up the FSA Web server.

3.4 Contact

For additional information on this feasibility report, please contact Sheila A. Suppinsky, Senior Systems Administrator, Federated Scientific Associates, at (513) 873-5124, Fax (513) 873-5009, or e-mail suppin@fsa_1.com.

4. Notes

[1] Federated Scientific Associates, *Coordinated Approved Vendor Catalog,* No. 42, Issue 14, June 2000.

[2] DataStuff International, Technical Specifications, www.datastuff.com/tech/31203.html, updated November 28, 2000.

[3] DataStuff International, Technical Specifications, www.datastuff.com/tech/33298.html, updated November 28, 2000.

5. Appendix 1

Federated Scientific

headquarters memorandum

DATE: July 15, 2000

TO: FSA/ISD

FROM: FSA/HQ (IS)

SUBJECT: Feasibility Report for the FSA Web Server's Backup Capability

We are concerned that, given the increased utilization and reliance on the FSA Web server, we do not have adequate backup capabilities should a catastrophic failure of the server's hard disk occur. Please provide FSA/HQ (IS) a Feasibility Report, including your analysis and recommendation for redressing this deficiency NLT August 31, 2000.

Feasibility Report Checklist

- Have I defined the problem to the point where the requirements for solving it are clear?
- Have I selected a manageable number of candidate solutions that are apparently viable?
- Have I developed criteria that relate to the problem?
- Have I explained all criteria, including why they were selected and how much weight each is being given?
- For all criteria, have I collected information (data) that is objective and meaningful?

- Have I provided useful interpretations of this information (data)?
- Have I included a conclusion based on these interpretations?
- Have I made a recommendation based on this conclusion?
- Have I included a contact who can provide more information about this feasibility report?
- Have I documented the sources I used for my information?
- Have I included in the appendix any necessary supporting documents?

8

Instructions and Manuals

Explaining to someone how to do something can be a challenging task. If you do not believe that, try showing a young child how to tie his or her shoes. Both instructions and manuals describe processes so that readers can accomplish the steps required to successfully complete a task. In technical writing, instructions deal with narrowly defined topics where the goal is to explain how to do something and not much more. Broader requirements demand complex descriptions in the form of manuals. Both instructions and manuals have the same objective, but manuals contain more than just instructions for carrying out a simple task. Fundamentally, however, they both do the same thing and work the same way.

What Are Instructions?

Instructions are process descriptions for human involvement. You probably remember that process descriptions were discussed in Chapter 4. The process description is a good starting point for writing instructions; but instructions involve more than simply describing a process. When you give instructions, you are not only describing the steps of a process, you are also describing how to accomplish these steps. That means you are accepting the additional responsibility of describing the process accurately and in a way that your reader can follow safely and effectively. It is the difference between describing to a medical student what happens when an inflamed appendix is removed, and describing to a surgical resident, hovering over a patient with scalpel in hand, how to remove an inflamed appendix. To appreciate the difference, just imagine that you are the patient.

As Outline 8.1 shows, providing technical instructions is actually straightforward if you understand the process, as well as the specific needs and skill level of your audience.

This chapter will illustrate how to use Outline 8.1 by giving instructions for executing the most powerful of all martial arts techniques, the FinkelKick. These instructions will show the reader how to smash a reinforced concrete wall with his or her bare foot. Instructions often carry disclaimers because of potential risks and liabili-

---

Outline 8.1 Instructions

Introduction
- Define the overall process.
- Describe its purpose.
- Explain any needed theory or principles.
- List the steps.

Discussion
- For each step listed above:
  - Define the step.
  - Describe generally what happens in this step.
  - Provide needed information specific to this step:
    - Note any dangers and cautions.
    - List required equipment or tools.
  - Provide specific directions for executing this step.
  - Describe the result that should occur.
  - Transition to the next step (if there is one).

Conclusion
- Briefly summarize the steps of the process.
- Tell the reader where to find additional information (if applicable).

ties, so it should be no surprise that these instructions have the following disclaimer:

> Only certifiable fools who are capable of believing anything should attempt the following technique. Attacking a reinforced concrete wall with the FinkelKick will result in severe orthopedic damage and unimaginable pain and suffering. The fact that one even attempts this fictitious technique can and will be used as evidence in his or her commitment hearing.

Introduction

Following Outline 8.1, begin with a definition of the process and a description of the purpose:

> In martial arts, the FinkelKick is a specialized foot technique by which relatively large pressures are developed through the application of substantial kinetic energy to a limited target area. The Finkel-Kick is commonly used as a power technique in competitive, demonstration events.

Next, discuss any theory or principles needed by the reader to understand how to do the FinkelKick. Obviously, using this kick to break a concrete wall would rely more on the reader's low cognitive complexity than any understanding of theory or principles; however, here is some theory anyway just for the sake of exemplification:

> The power of the FinkelKick depends on kinetic energy, which is the energy a martial artist possesses by virtue of his or her motion. Kinetic energy (ke) depends on the mass (m) of the martial artist and the velocity (v) at which this mass is moving (Equation 1).

$$ke = 1/2mv^2 \qquad (1)$$

> To fully exploit the role of velocity in generating kinetic energy, the martial artist executes the FinkelKick with the greatest possible velocity of his

or her mass (m), resulting in high rates of deceleration (a) when contacting the wall. This technique can generate a very large force (f) (Equation 2).

$$f = ma \qquad (2)$$

This force (f) is then applied to a limited area (A) with the ball of the foot to create the kick's relatively large pressure (P) (Equation 3).

$$P = f/A \qquad (3)$$

Now complete the introduction by listing the steps required:

Accomplishing the FinkelKick requires the execution of five steps: positioning, chambering, extending, retracting, and resetting.

Discussion

At this point, provide specific instructions for each step. Start with Step 1, positioning. Notice that the actual instructions are written in a snappy, imperative style where the subject, "you," is understood. Describing Step 1 involves defining the step, overviewing what happens in the step, providing necessary information such as danger and caution notices, and noting any equipment required.

Step #1: Positioning

Positioning is the preparatory action of placing yourself in front of one of the long sides of the concrete wall. It does not matter which side.

In this step you will assume a traditional front stance about three feet from the target. Make sure that your feet are about shoulder width apart and that your right foot is about two feet in front of your left foot. The toes on both feet should be facing forward. Your front leg should be bent while the back leg should be straight.

Caution: Ensure that the ground under your feet is dry, level, and free of sharp objects.

Equipment: You will need a reinforced concrete wall and a martial arts uniform.

Next provide the specific execution instructions, describe the result, and transition to the next step:

Assume a front stance about three feet in front of the wall.

You should be facing the wall, with the toes of your right foot about three feet from the wall. Once in the correct position, you are ready for chambering.

Step #2: Chambering

First define the step; then describe what happens, provide necessary information, and note any equipment required:

Chambering is a posturing action by which you position the right (kicking) leg for the attack. A good chamber is essential if the kick is to contact the wall at the highest possible velocity and at the desired angle of 90 degrees.

In this step you will raise your right knee as high as possible and then hold your leg in that position with your toes still pointing forward toward the wall. You will also extend your hands toward the wall.

Caution: Ensure that your back is straight and perpendicular to the ground before assuming the chambered position; otherwise, you will fall.

Equipment: Same as in Step #1.

Next provide the specific execution instructions, describe the result, and transition to the next step:

Raise your right knee as high as possible while extending both hands toward the wall.

You should be balancing on your left foot, with your back straight and perpendicular to the ground and with both hands extended toward the wall.

Note: Your right hand may be only a few inches from the wall's surface. Once in this position, you are ready to extend your foot into the wall.

Step #3: Extending

As before, define the step, overview what happens, provide necessary information, and list any equipment required. Note the addition of a "danger flag" for this step.

Extending is the front snapping action that drives the foot into the wall.

In this step you will thrust the ball of the foot through the wall while raising your hands up and to the side of your body. Your kicking leg should be at a 90-degree angle to the wall at the time of contact. Remember: You are kicking a structure that consists of tons of jagged concrete and rusty reinforcing steel members. So keep your mental focus.

Caution: Make sure you hit the wall with the ball of your foot and not your toes.

Danger: To prevent injury from flying debris, ensure that all spectators and support personnel are 50 yards from the wall.

Equipment: Same as Steps #1 and #2. The only additional equipment required for this step is an ambulance staffed with paramedics.

Next provide the specific execution instructions, describe the result, and transition to the next step:

Mentally focus deep into the wall, then accelerate your foot straight out to that point of focus as you raise your hands up and to the side of your body.

If all goes well, you should hear sounds of the wall's structure failing catastrophically, and you should see multiple fault cracks rapidly forming. If all does *not* go well, you should hear sounds of your foot's skeletal structure failing catastrophically, and you should feel intense pain and regret. In either case, you are ready to retract your foot from the wall.

Step #4: Retracting

Again, define the step, describe what happens, provide necessary information, and note any equipment required:

> *Retracting* is a deposturing action by which you pull your foot from the wall back into a chambered position. Retracting is normally done in competitive sparring to prevent your opponent from grabbing your leg. In this wall-breaking demonstration it is a matter of good form.
>
> In this step, you will simply pull your foot back from the broken wall and raise your knee as high as possible while pulling your arms back to the chambered position. Again, your toes (or what is left of them) should be pointing toward the wall, and your hands should be in front of your body and extended straight at the target.
>
> Caution: If your foot is hemorrhaging badly, forget proper form and stop the bleeding.
>
> Equipment: Same as Step #3.

As before, provide the specific execution instructions, describe the result, and transition to the next step:

> Pull your kicking foot back toward your body, raising your right knee as high as possible while extending your arms in front of you toward the destroyed wall.
>
> You now should be facing what is left of the wall, with your right knee raised and both hands extended toward the destroyed wall. Your right hand may be within a few inches of where the wall once stood. You are now ready for the final step of resetting.

Step #5: Resetting

Define the step, describe what happens, provide additional information, and note any equipment required:

Resetting is the final action of placing yourself into your initial, starting position.

In this step you will reassume a traditional front stance about three feet from where the wall used to be. Again, make sure that your feet are about shoulder width apart and that your right foot is about two feet in front of your left foot. The toes (or what is left of them) should be facing forward. Your front leg should be bent, while the back leg should be straight.

Caution: Arrange for immediate psychiatric care if you have not already done so.

Equipment: Same as Step #3.

Finally, provide the specific execution instructions and describe the result. Since this is the last step, no transition is required.

Lower your right leg from its chambered position and assume a front stance.

Your right leg should be forward and bent, while your left leg should be to the rear and straight. You should be facing the wall, with your right foot about three feet from where the wall once stood. Once in this position, you have successfully completed the FinkelKick.

Conclusion

Briefly summarize the process and tell your reader where to find additional information:

Execution of the FinkelKick requires five steps: positioning, chambering, extending, retracting, and resetting. Together these steps result in a high–kinetic energy technique capable of destroying a reinforced concrete wall.

For additional information on the scientific basis of the FinkelKick, consult any good physics textbook for discussions of kinetic energy, force, and pressure.

Putting It All Together   Here is the complete instruction set for the FinkelKick, including fully integrated visuals.

Note the use of layout and font formatting, as well as white space, to enhance the presentation of the information.

Breaking a Reinforced Concrete Wall with the FinkelKick Martial Arts Technique

Disclaimer

Only certifiable fools who are capable of believing anything should attempt the following technique. Attacking a reinforced concrete wall with the FinkelKick will result in severe orthopedic damage and unimaginable pain and suffering. The fact that one even attempts this fictitious technique can and will be used as evidence in his or her commitment hearing.

Introduction

In martial arts, the FinkelKick is a specialized foot technique by which relatively large pressures are developed through the application of substantial kinetic energy to a limited target area. The FinkelKick is commonly used as a power technique in competitive, demonstration events.

The power of the FinkelKick depends on kinetic energy, which is the energy a martial artist possesses by virtue of his or her motion. Kinetic energy (ke) depends on the mass (m) of the martial artist and the velocity (v) at which this mass is moving (Equation 1).

$$ke = 1/2mv^2 \qquad (1)$$

To fully exploit the role of velocity in generating kinetic energy, the martial artist executes the FinkelKick with the greatest possible velocity of his or her mass (m), resulting in high rates of deceleration (a) when contacting the wall. This technique can generate a very large force (f) (Equation 2).

$$f = ma \qquad (2)$$

This force (f) is then applied to a limited area (A) with the ball of the foot to create the kick's relatively large pressure (P) (Equation 3).

$$P = f/A \qquad (3)$$

Accomplishing the FinkelKick requires the execution of five steps: positioning, chambering, extending, retracting, and resetting.

Discussion
Step #1: Positioning

Positioning is the preparatory action of placing yourself in front of one of the long sides of the concrete wall. It does not matter which side. In this step you will assume a traditional front stance about three feet from the target. Make sure that your feet are about shoulder width apart and that your right foot is about two feet in front of your left foot. The toes on both feet should be facing forward. Your front leg should be bent while the back leg should be straight.

Caution

Ensure that the ground under your feet is dry, level, and free of sharp objects.
Equipment: You will need a reinforced concrete wall and a martial arts uniform.

1. Assume a front stance about three feet in front of the wall.

You should be facing the wall, with the toes of your right foot about three feet from the wall. See Figure 8.1. Once in the correct position, you are ready for chambering.

Step #2: Chambering

Chambering is a posturing action by which you position the right (kicking) leg for the attack. A

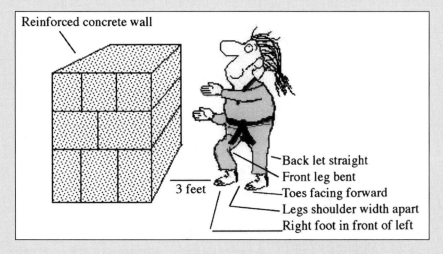

Reinforced concrete wall

3 feet

Back let straight
Front leg bent
Toes facing forward
Legs shoulder width apart
Right foot in front of left

Figure 8.1
Positioning

good chamber is essential if the kick is to contact the wall at the highest possible velocity and at the desired angle of 90 degrees. In this step you will raise your right knee as high as possible and then hold your leg in that position with your toes still pointing forward toward the wall. You will also extend your hands toward the wall.

Caution

Ensure that your back is straight and perpendicular to the ground before assuming the chambered position; otherwise, you will fall.
Equipment: Same as Step #1.

2. Raise your right knee as high as possible while extending both hands toward the wall.
You should be balancing on your left foot, with your back straight and perpendicular to the ground and with both hands extended toward the wall. See Figure 8.2. Note: Your right hand may be only a few inches from the wall's surface. Once in this position, you are ready to extend your foot into the wall.

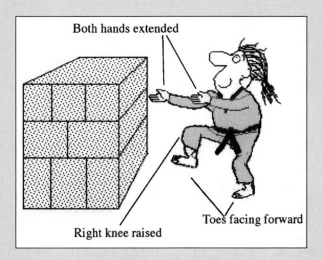

Both hands extended

Right knee raised

Toes facing forward

Figure 8.2
Chambering

Step #3: Extending

Extending is the front snapping action that drives your foot into the wall. In this step you will thrust the ball of the foot through the wall while raising your hands up and to the side of your body. Your kicking leg should be at a 90-degree angle to the wall at the time of contact. Remember: You are kicking a structure that consists of tons of jagged concrete and rusty reinforcing steel members. So keep your mental focus.

Caution

Make sure you hit the wall with the ball of your foot and not your toes.

Danger

To prevent injury from flying debris, ensure that all spectators and support personnel are 50 yards from the wall.
Equipment: Same as Steps #1 and #2. The only additional equipment required for this step is an ambulance staffed with paramedics.

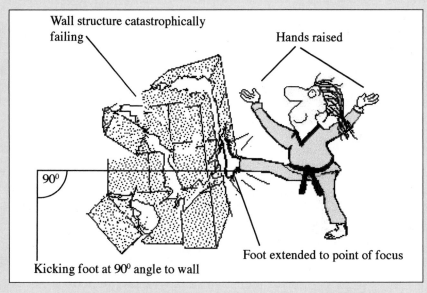

Wall structure catastrophically failing

Hands raised

90^0

Kicking foot at 90^0 angle to wall

Foot extended to point of focus

Figure 8.3
Extending

3. Mentally focus deep into the wall, then accelerate your foot straight out to that point of focus as you raise your hands up and to the side of your body.
If all goes well, you should hear sounds of the wall's structure failing catastrophically, and you should see multiple fault cracks rapidly forming. See Figure 8.3. If all does *not* go well, you should hear sounds of your foot's skeletal structure failing catastrophically, and you should feel intense pain and regret. In either case, you are ready to retract your foot from the wall.

Step #4: Retracting

Retracting is a deposturing action by which you pull your foot from the wall back into a chambered position. Retracting is normally done in competitive sparring to prevent your opponent from grabbing your leg. In this wall-breaking demonstration it is a matter of good form. In this step, you will simply pull your foot back from the broken wall and raise your knee as high as possible while pulling your arms back to the chambered position. Again, your toes (or what is left of them) should be pointing toward the wall, and your hands should be in front of your body and extended straight at the target.

Caution

If your foot is hemorrhaging badly, forget proper form and stop the bleeding.
Equipment: Same as Step #3.

4. Pull your kicking foot back toward your body, raising your right knee as high as possible while extending your arms in front of you toward the destroyed wall.
You now should be facing what is left of the wall with your right knee raised, and both hands extended toward the destroyed wall. See Figure 8.4. Your right hand may be within a few inches of where the wall once stood. You are now ready for the final step of resetting.

Figure 8.4
Retracting

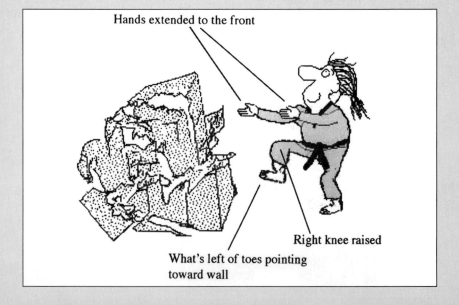

Hands extended to the front

Right knee raised

What's left of toes pointing
toward wall

Step #5: Resetting

Resetting is the final action of placing yourself into your initial, starting position. In this step you will reassume a traditional front stance about three feet from where the wall used to be. Again, make sure that your feet are about shoulder width apart and that your right foot is about two feet in front of your left foot. The toes (or what is left of them) should be facing forward. Your front leg should be bent, while the back leg should be straight.

Caution

Arrange for immediate psychiatric care if you have not already done so.
Equipment: Same as Step #3.

5. Lower your right leg from its chambered position and assume a front stance.

Your right leg should be forward and bent, while your left leg should be to the rear and straight. You should be facing the wall, with your right foot about three feet from where the wall once stood. See Figure 8.5. Once in this position, you have successfully completed the FinkelKick.

Figure 8.5
Resetting

Front leg bent

Wall destroyed

Back leg straight

Conclusion

Execution of the FinkelKick requires five steps: positioning, chambering, extending, retracting, and resetting. Together these steps result in a high–kinetic energy technique capable of destroying a reinforced concrete wall.

For additional information on the scientific basis of the FinkelKick, consult any good physics textbook for discussions of kinetic energy, force, and pressure.

More on Manuals

As mentioned earlier, manuals follow the same basic pattern as instructions for telling someone how to do something. However, manuals have a much larger scope and are more comprehensive.

Manuals provide instructions for many complicated tasks involving complex equipment. In some cases, where systems are large and extremely complex (such as an aircraft or automobile), many separate manuals are required and may, in fact, fill a small library. Smaller systems (such as a lawnmower) frequently require only a single, self-contained manual.

Here are the most common types of manuals and the kinds of tasks with which they are used:

- *Assembly manuals:* construction, alignment, calibration, testing, or adjusting a mechanism.
- *Owner's manuals:* use of a mechanism, routine maintenance, and basic operation.
- *Operator manuals:* use of a mechanism and very minor maintenance.
- *Service manuals:* routine maintenance of a mechanism, including troubleshooting, testing, repairing, or replacing defective parts.
- *Technical manuals:* parts specifications, operation, calibration, alignment, diagnosis, and assembly.

- Do I understand the process and the skill level of the intended audience?

- Have I defined the overall process and described its purpose?
- Have I explained any needed theories or principles?
- Have I listed the steps?
- For each step:
  - Have I defined the step?
  - *Before* telling the reader to actually do something . . .
    - Have I given an overview of what will happen in this step?
    - Have I provided needed information such as cautions, dangers, and required equipment?
  - *When* telling the reader to do something . . .
    - Have I described exactly what should be done?
    - Have I described the result that should occur?
    - Have I transitioned to the next step if there is one?
- Have I summarized all the steps?
- Have I told the reader how to get additional information?

9

Laboratory and Project Reports

Remember those old horror movies in which the insane scientist transplanted human brains from one body to another in the laboratory? The labs were always impressive. All around one could see and hear high-voltage arcs traveling up Jacob's ladders and corona discharges zapping off spherical electrodes. What those movies did not show was the scientist typing up a laboratory or project report after the experiments were concluded. One does not successfully transplant a brain from one body to another and not write a report!

Virtually anyone working in engineering and the sciences will be tasked, in one fashion or another, with producing a laboratory or project report. These documents present information that relates to the controlled testing of a hypothesis, theory, or device using test equipment (the *apparatus*) and a specified series of steps employed to perform the test (the *procedure*). The emphasis in a laboratory or project report is on documenting the design and conduct of the test, how the variables were controlled, and what the resulting data show.

**What Are
Laboratory
and Project
Reports?**

In their purest form, *laboratory reports* are research-oriented documents, meaning that they start with a hypothesis or theory that needs to be tested under highly controlled conditions. For example, suppose you are an aeronautical engineer hypothesizing that your new wing design could be used to generate higher lift at hypersonic speeds with increased flight stability. To test that hypothesis in a laboratory, you would need an apparatus—in this case, a hypersonic wind tunnel and a model of your wing design. You would also need a procedure for using that wind tunnel to test your wing design model. You could then use the procedure to collect data from the wind tunnel tests and interpret the data to see if your new wing design generated higher lift with increased stability under hypersonic conditions. Finally, you could assess whether the original hypothesis was supported and, if so, probably recommend that more research be done.

Laboratory reports can also take the form of *project reports,* which are commonly used in teaching laboratories. Instead of testing a hypothesis or theory, you would focus on fulfilling specific requirements that normally would come from a teacher in the form of a project assignment. The goal of the project might be to demonstrate the application of a theory or set of theories using available technology. For example, you and several other students in an engineering project course might be asked to build a motorized vehicle with a homeostatic control system that would track accurately along a 100-yard course. The course could be marked with a white line and could contain six 90-degree curves. Your group's vehicle also would have to avoid several obstacles placed in its path along the way while completing the course within a specified time.

As students, you would design and build an operating vehicle within the cost and equipment constraints prescribed by the teacher. Your group would probably apply basic control theory using optical and ultrasonic sensors, several electromagnetic servos, and an embedded microprocessor. In effect, your vehicle and the test track would be the apparatus, and the process used to test your vehicle's performance would be the procedure. As part of the class, your group would make several test runs of the vehicle, collect data relating to speed and accuracy, and then develop a project report to document how well your vehicle met the requirement.

Both laboratory and project reports, then, follow the same general pattern, as shown in Outline 9.1.

For illustration, we will use Outline 9.1 to develop a laboratory report. A project report

---

Outline 9.1 Laboratory or Project Report

Introduction
- Purpose   What is the purpose of this report?
- Problem   What is the hypothesis or requirement?
- Scope   What are the limitations of this report?

Background
- Theory   What is the theoretical basis of this project?
- Research   What prior research is relevant to this project?

Test and Evaluation
- Apparatus   What device(s) did you use?
- Procedure   What procedure(s) did you use?

Findings
- Data   What were the results of the test?

Interpretation   What is your interpretation of the results?

Conclusion   What can you conclude from the interpretation(s)?
- Recommendations   What is your recommendation based on this conclusion?

would be similar, but would replace the hypothesis with a specific problem to be solved; it would focus on how well the project was accomplished rather than on how well the hypothesis was supported. Otherwise, the format would be the same for both reports.

This laboratory report will focus on the efficacy of using the FinkelTUBE beam power pentode electron tube in power audio amplifier applications. As you might imagine, the FinkelTUBE is a powerful piece of vaporware technology that can be found only in this book.

Introduction

First, in the introduction, describe the purpose of the report:

Purpose
This laboratory report documents the preliminary test and analysis accomplished on the FinkelTUBE beam power pentode to determine its effectiveness when used as an audio frequency power amplifier in a demanding environment like a rock concert.

Next provide a statement of the problem requiring this report:

Problem
SoundProducts International, in its memorandum of August 8, 2000, has noted the increasing application of electron tube, power amplifiers in demanding audio environments. A wide range of modern tube amplifiers is now capturing an increasing share of the high-end amplifier market. As part of its marketing assessment of the device as a potential audio amplifier product, SoundProducts needs a preliminary laboratory analysis of the FinkelTUBE beam power pentode for use in audio, power amplifier applications.

Finally, round out the introduction with a brief statement of the scope of the report:

Scope

This preliminary test and evaluation of the Finkel-TUBE in a power audio amplifier application includes only an analysis of power, distortion, and noise. Furthermore, this report is limited to the technical efficacy of the device and does not include analysis of its cost–benefit ratio or marketability in audio applications.

Background

In the background section, provide the information necessary for the reader to understand and appreciate the test report and findings that will follow. This section should review any relevant theory and past research that the reader needs.

Theory

Vacuum tubes were first developed in the early 1900s to control the flow of electronic current and were used in virtually all audio amplifier equipment until the early 1960s. At that time, solid-state devices quickly replaced tubes in audio applications because of their smaller size and higher efficiencies.

In recent years, however, there has been a noted resurgence in audio amplifier applications using vacuum tubes. Apocryphal data indicate that many audiophiles, audio engineers, and professional musicians prefer the tube amplifier's "warm and softer" sounds. Some attribute the tube's comparatively better sound, vis-à-vis solid state, to some or all of the following: better power management, less distortion, improved signal-to-noise ratios, better linearity, less clipping, and fewer feedback problems. This preliminary analysis focuses on the first three variables of power, distortion, and noise.

The FinkelTUBE is a beam power pentode designed primarily as a wide-band, radio frequency (RF) power amplifier running in Class C mode across the 100 KHz–200 MHz spectrum. This research is designed to determine the technical efficacy of using the FinkelTUBE as an audio frequency (AF) power amplifier running in Class A across the DC–20 KHz spectrum.

Likewise, review any research that is relevant to the inquiry at hand.

Research
Limited research has been done on using RF power tubes in AF applications. The company's Advanced Products Division did investigate the limited use of low-powered transmitting tubes (primarily the 6146) in audio applications (Williams 1990, 347–350). This investigation indicated that, while the tube could function effectively at AF frequencies and provide solid performance at output powers up to 50 watts, the cost factors were not favorable for such applications. Of course, this analysis predated the current resurgence in tube amplifier applications, as well as the state-of-the-art technology capabilities of the FinkelTUBE.

Test and Evaluation
In the test and evaluation section, describe the apparatus and procedure used to do this research.

Apparatus
The FinkelTUBE is capable of producing a continuous output of 5,000 watts of audio into an 8-ohm load with an input signal of 10 volts peak-to-peak. This power is more than adequate for demanding environments such as rock concerts and should easily achieve the apparent goal of destroying the high-frequency hearing of anyone who attends.

Local zoning ordinances and environmental protection considerations would not permit us to evaluate the FinkelTUBE at its maximum rating in the laboratory. Consequently, for test purposes, the FinkelTUBE was used in a single-ended audio amplifier, running in Class A, in a device designated the FinkelAMP. The *FinkelAMP* uses a multistage, high-gain voltage amplifier driving the FinkelTUBE power amplifier with a continuous output power of 100 watts.

The unit was built on a steel chassis using traditional sockets and point-to-point wiring. The chassis was enclosed in a wooden cabinet with a Plexiglas window that provides a view port for various instrumentation indicators, including signal lights, and a multifunction meter reading plate and grid voltages and currents. The amplifier was driven by a calibrated signal generator, and the output was fed into a purely resistive 8-ohm load, where it was analyzed by wide-bandwidth, multitrace oscilloscopes, as well as high-end computers running proprietary analysis software.

Procedure

A pure, distortion-free, sine-wave signal of 1 volt peak-to-peak was applied to the voltage amplifier stages of the FinkelAMP. This level was the maximum input signal amplitude possible without saturating these stages. The frequency of the signal was varied across a range from DC to 20 KHz in 1 KHz increments. The output was analyzed in terms of power management, distortion, and noise.

Findings

Discuss the data that the tests yielded. As you will see, when this paper is assembled with visuals later in this chapter, it will include a photograph of the FinkelAMP test bench, as well as several visuals that display the results of the tests. For now, however, just focus on discussing the results.

Data

The data gathered fell into these three categories: power in watts, distortion as a percentage of total signal, and signal-to-noise ratio.

Power

Power is the product of voltage and current output from the amplifier. Constancy of power across the full frequency range is essential for properly driving complex speaker systems in demanding environments. The power output of the FinkelAMP was observed across the full range of frequencies tested.

Although the output dropped off slightly above 18 KHz, overall, the output remained relatively constant. This constancy was especially true in the primary audio frequencies between DC and 15 KHz.

Distortion
Distortion is the unwanted characteristic of an amplifier to modify the nature of a signal while amplifying it. Although solid-state amplifiers are often credited with having lower distortion values than tube amplifiers, the FinkelAMP exhibited harmonic distortion levels across its total operating range of less than .05, which is as good as, if not better than, most solid-state amplifiers. The distortion that did occur was at the upper limit of the range, well above the threshold of human hearing.

Noise
Noise includes spurious output signals that were not part of the original input signal. Good solid-state amplifiers typically have a signal-to-noise ratio (SN) of .15 mV. By comparison, the Finkel-AMP provided a typical SN of .09–.15 across the primary audio spectrum.

Interpretation

Provide your interpretation of the data. Do not come to any overall conclusion in this section.

Interpretation
In terms of the preliminary tests involving power, distortion, and noise, the FinkelTUBE-based Finkel-AMP performed well in an audio frequency, power amplifier mode. Although the high-power levels could not be duplicated in the laboratory setting, the low-power tests demonstrated excellent results.

Conclusion

Finally, provide your overall conclusions relating to the original purpose of this study, and make any specific recommendations that you believe are in order.

Conclusion
Initial laboratory analysis, based on the Finkel-TUBE's measured performance, indicates that technically it is a viable candidate for high-power, audio, power amplifier applications.

Recommendation
SoundProducts should conduct a more complete laboratory analysis involving the following tests: high-power management, clipping, feedback, harmonics, and linearity. Assuming good results from these tests, SoundProducts should consider the FinkelTUBE a technically efficacious option for audio, power amplifier applications.

Here is the assembled laboratory report, complete with photographs, tables, and charts.

Putting It All Together

Laboratory Report
The FinkelTUBE in Power Audio Applications

Submitted to: Mr. James E. Carson, President
SoundProducts International
1 Town Plaza
New York, NY 12002

Submitted by: Dr. Robert W. Yinburg, Chief Scientist
FinkelLaboratories, Incorporated
3 Prestige Way
New York, NY 12034

September 30, 2000

Introduction
Purpose
This laboratory report documents the preliminary test and analysis accomplished on the Finkel-TUBE beam power pentode to determine its effectiveness when used as an audio frequency power amplifier in a demanding environment like a rock concert.

Problem

SoundProducts International, in its memorandum of August 8, 2000, has noted the increasing application of electron tube, power amplifiers in demanding audio environments. A wide range of modern tube amplifiers is now capturing an increasing share of the high-end amplifier market. As part of its marketing assessment of the device as a potential audio amplifier product, SoundProducts needs a preliminary laboratory analysis of the FinkelTUBE beam power pentode for use in audio power amplifier applications.

Scope

This preliminary test and evaluation of the FinkelTUBE in a power audio amplifier application includes only an analysis of power, distortion, and noise. Furthermore, this report is limited to the technical efficacy of the device and does not include analysis of its cost–benefit ratio or marketability in audio applications.

Background

Theory

Vacuum tubes were first developed in the early 1900s to control the flow of electronic current and were used in virtually all audio amplifier equipment until the early 1960s. At that time, solid-state devices quickly replaced tubes in audio applications because of their smaller size and higher efficiencies.

In recent years, however, there has been a noted resurgence in audio amplifier applications using vacuum tubes. Apocryphal data indicate that many audiophiles, audio engineers, and professional musicians prefer the tube amplifier's "warm and softer" sounds. Some attribute the tube's comparatively better sound, vis-à-vis solid state, to some or all of the following: better power management, less distortion, improved signal-to-noise ratios, better linearity, less clipping, and fewer feedback problems. This preliminary analysis focuses on the first three variables of power, distortion, and noise.

The FinkelTUBE is a beam power pentode designed primarily as a wide-band, radio frequency (RF) power amplifier running in Class C mode across the 100 KHz–200 MHz spectrum. This research is designed to determine the technical efficacy of using the FinkelTUBE as an audio frequency (AF) power amplifier running in Class A across the DC–20 KHz spectrum.

Research

Limited research has been done on using RF power tubes in AF applications. The company's Advanced Products Division did investigate the limited use of low-powered transmitting tubes (primarily the 6146) in audio applications (Williams 1990, 347–350). This investigation indicated that while the tube could function effectively at AF frequencies and provide solid performance at output powers up to 50 watts, the cost factors were not favorable for such applications. Of course, this analysis predated the current resurgence in tube amplifier applications, as well as the state-of-the-art technology capabilities of the FinkelTUBE.

Test and Evaluation

Apparatus

The FinkelTUBE is capable of producing a continuous output of 5,000 watts of audio into an 8-ohm load with an input signal of 10 volts peak-to-peak. This power is more than adequate for demanding environments such as rock concerts and should easily achieve the apparent goal of destroying the high-frequency hearing of anyone who attends.

Local zoning ordinances and environmental protection considerations would not permit us to evaluate the FinkelTUBE at its maximum rating in the laboratory. Consequently, for test purposes, the FinkelTUBE was used in a single-ended audio amplifier, running in Class A, in a device designated the FinkelAMP. The *Finkel-AMP* uses a multistage, high-gain voltage amplifier driving the FinkelTUBE power amplifier with a continuous output power of 100 watts.

The unit was built on a steel chassis using traditional sockets and point-to-point wiring. The chassis was enclosed in a wooden cabinet with a Plexiglas window that provides a view port for various instrumentation indicators, including signal lights, and a multifunction meter reading plate and grid voltages and currents. The amplifier was driven by a calibrated signal generator, and the output was fed into a purely resistive 8-ohm load, where it was analyzed by wide-bandwidth, multitrace oscilloscopes, as well as high-end computers running proprietary analysis software. See Photo 9.1.

Photo 9.1
FinkelAMP
laboratory test
bench

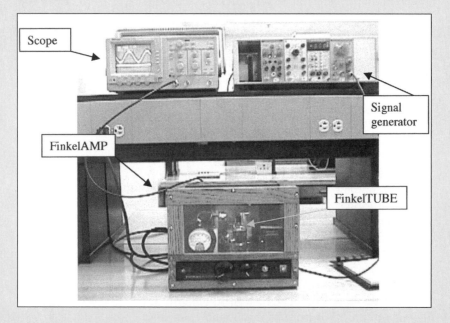

Procedure

A pure, distortion-free, sine-wave signal of 1 volt peak-to-peak was applied to the voltage amplifier stages of the FinkelAMP. This level was the maximum input signal amplitude possible without saturating these stages. The frequency of the signal was varied across a range from DC to 20 KHz in 1 KHz increments. The output was analyzed in terms of power management, distortion, and noise.

Findings

Data

The data gathered fell into these three categories: power in watts, distortion as a percentage of total signal, and signal-to-noise ratio.

Power. Power is the product of voltage and current output from the amplifier. Constancy of power across the full frequency range is essential for properly driving complex speaker systems in demanding environments. The power output of the FinkelAMP was observed across the full range of frequencies tested. Although the output dropped off slightly above 18 KHz, overall, the output remained relatively constant. This constancy was especially true in the primary audio frequencies between DC and 15 KHz. (See Figure 9.1.)

Distortion. Distortion is the unwanted characteristic of an amplifier to modify the nature of a signal while amplifying it. Although solid-state amplifiers are often credited with having lower distortion values than tube amplifiers, the FinkelAMP exhibited harmonic distortion levels across its total operating range of less than .05, which is as good as, if not better than, most solid-state amplifiers. The distortion that did occur was at the upper limit of the range, well above the threshold of human hearing. (See Figure 9.2.)

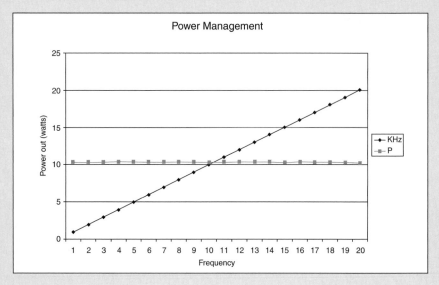

Figure 9.1
Power

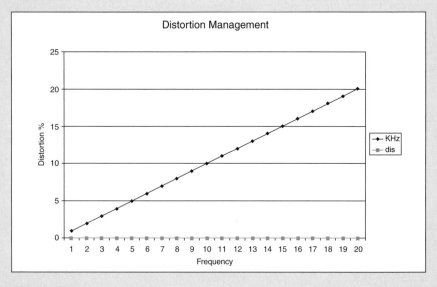

Figure 9.2
Harmonic distortion

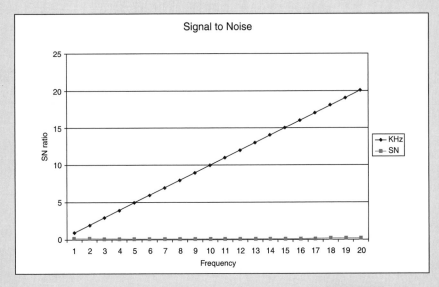

Figure 9.3
Signal to noise

Noise. Noise includes spurious output signals that were not part of the original input signal. Good solid-state amplifiers typically have a signal-to-noise ratio (SN) of .15 mV. By comparison, the FinkelAMP provided a typical SN of .09–.15 across the primary audio spectrum. (See Figure 9.3.)

Interpretation
In terms of the preliminary tests involving power, distortion, and noise, the FinkelTUBE-based FinkelAMP performed well in an audio frequency, power amplifier mode. Although the high-power levels could not be duplicated in the laboratory setting, the low-power tests demonstrated excellent results. (See Table 9.1.)

Conclusion
Initial laboratory analysis, based on the FinkelTUBE's measured performance, indicates that technically it is a viable candidate for high-power, audio, power amplifier applications.

Table 9.1 Summary of results

KHz	P	dis	SN
1	10.22	.01	.15
2	10.24	.01	.15
3	10.26	.01	.10
4	10.30	.01	.10
5	10.29	.01	.09
6	10.29	.01	.09
7	10.30	.01	.09
8	10.30	.01	.09
9	10.29	.01	.10
10	10.29	.01	.10
11	10.31	.01	.10
12	10.30	.01	.11
13	10.31	.01	.11
14	10.33	.01	.11
15	10.21	.01	.12
16	10.30	.02	.12
17	10.29	.01	.13
18	10.00	.01	.14
19	10.00	.01	.14
20	10.00	.01	.15

KHz = frequency in kilohertz;
P = power in watts; dis = distortion in
percent; SN = signal-to-noise ratio.

Recommendation

SoundProducts should conduct a more complete laboratory analysis involving the following tests: high-power management, clipping, feedback, harmonics, and linearity. Assuming good results from these tests, SoundProducts should consider the FinkelTUBE a technically efficacious option for audio, power amplifier applications.

Source

Robert E. Williams. "Investigation of Transmitting Tubes for Audio Applications," *Technical Tidbits*. SoundProducts International, 1990, pp. 345–354.

- Have I clearly defined the purpose of this report?
- Have I clearly described the problem that requires this report?
- Have I clearly explained the limitations of this report?
- Have I discussed any theory necessary for the reader to understand the report?
- Have I reviewed relevant prior research?
- Have I described the apparatus I used to collect the data?
- Have I described the procedure I used to collect the data?
- Have I clearly provided the results of the test?
- Have I properly interpreted these results?
- Have I made proper conclusions from these interpretations?
- Have I made a recommendation based on my conclusions?

Laboratory and Project Report Checklist

10

Research
Reports

Research Reports are similar to those research papers that every student has done at one time or another. You remember—papers with titles like "The Sexuality of Lady Macbeth" or "The Benefits of the Modern Mosquito." In technical writing, however, research reports are focused, objective inquiries into technical subjects.

Research reports describe the discovery, analysis, and documentation of knowledge through some type of investigation. In technical writing they are specifically geared to the purpose at hand, the readers who will use them, and whatever scope limitations exist. Technical research reports frequently focus on new, evolving, sometimes purely hypothetical technologies, in which case they are called *state-of-the-art-reports*.

What Are Research Reports?

One distinguishing characteristic of research reports is the extensive research and documentation required. The actual research may consist of library and laboratory research, interviews, questionnaires, various types of corporate technical reports, and trade journal articles. Also, research report writers increasingly use the wealth of information on the Internet.

The organization of a research report is straightforward, as shown in Outline 10.1. However, exactly what goes in the discussion

---

Outline 10.1 Research Report

Introduction	
• Purpose	Purpose of this report
• Problem	Brief overview of, or introduction to, the topic
• Scope	Limitations of this report
Background	
• Theory	Theoretical basis for understanding the topic (if needed)
• History	Historical perspective of the topic (if needed)
Discussion	Main section of the report
Conclusion	
• Summary	Summary of the discussion section
• Recommendation	Suggestions based on the summary
References	
• Sources cited	Sources consulted and used in the report
• Sources not cited	Sources consulted but not specifically used
Appendixes	Additional supporting material not needed to understand the report

---

section depends on the topic being researched and the specific requirements for the research. If the focus is on how we got to where we are in developing a certain technology, the discussion will be primarily historical. On the other hand, if the purpose is to research new, evolving technologies, the discussion may be geared more to future implications.

Research reports are usually comprehensive documents that often extend beyond the scope of this book. However, for illustration, this chapter will provide a truncated example, based on Outline 10.1, that contains the required elements of a research report.

Developing a Research Report

This example is a state-of-the-art report on the *FinkelCHIP Quantum Central Processing Unit (QCPU).* This powerful piece of vaporware exploits the tremendous potential of quantum computing architectures to effectively increase, by several orders of magnitude, the high-end computing power currently available. (Quantum computing is actually an area of serious scientific research among leading physicists and engineers. It is an extremely dynamic theoretical field.)

Introduction

As in other documents, the first part of the research report is the introduction section. Start with the purpose statement to explain why you are writing the report:

> *Purpose*
> The purpose of this report is to provide a state-of-the-art investigation of the FinkelCHIP Quantum Central Processing Unit (QCPU), including a theoretical review of the premises of its operation.

Next state the problem that the report addresses. In a research report, the problem is really more of a general background statement that expands on the topic and gives a brief context for what the report will investigate.

Note the inclusion of a source citation to support the assertion regarding processing speed. This type of documentation is required in research reports and is explained more thoroughly in Chapter 13.

> *Problem*
> Traditionally, computing power has been enhanced by increasing CPU speeds primarily through decreasing the size of conductors and solid-state devices used in chip fabrication. Decreasing size, however, has finite limitations, such as those associated with reducing the dielectric constants of the required materials. There has also been a move

toward increasing the number of instructions executed for each clock cycle, especially with reduced instruction set (RISC) processors.

FinkelProducts International has transcended this traditional paradigm by developing the QCPU. The QCPU exploits and manipulates the quantum nuclear spin states of atoms. The QCPU is capable of performing very large numbers of advanced computational tasks simultaneously using the superimposition of multiple values encoded into the respective spin states of individual atoms. The resulting (effective) CPU speed is equal to or better than 500 GHz (Josephson 1999, 291). This effective speed makes the QCPU an ideal chip for processing-intensive tasks such as cryptographic factoring of large numbers, DNA sequencing in genetic research, and interactive, three-dimensional holographic imaging in advanced virtual reality systems.

In any research paper, there is no way that you can possibly research everything about your topic. Human knowledge is not that simple or easy, and there is too much of it. So you will have to limit your paper by including only certain aspects of your topic. To complete the introduction, provide a scope statement that addresses this limitation. This section tells your reader what you are including in the paper and why; you will have to articulate some rationale for the limitations you are imposing.

Scope
The QCPU is built around proprietary information owned and protected by FinkelProducts International. Consequently, this report will be limited to the general theoretical approaches underlying the QCPU architecture; it will not investigate the actual methods used by the QCPU to implement these theoretical approaches.

Background

In the background section, discuss the theoretical and historical aspects of the topic, as appropriate. Given the purpose here, this example will focus only on the theoretical aspects of FinkelProducts' QCPU technology. The background should start with a brief discussion of quantum computing theory because this theory is not common knowledge for the audience; consequently, the theory discussion is essential to understanding the rest of the paper. Note again the inclusion of source citations throughout this discussion.

Theory

Quantum computing theory applies the knowledge of quantum physics to exploit subatomic phenomena of common elements to perform extremely complex computational tasks. When properly exploited, these phenomena provide a truly unprecedented ability for massively parallel processing (Ardvark 1999, 446–448). Several options exist to exploit quantum phenomena in this regard. One is to equate binary values to the ground and excited states of an atom. Another is to use traditional nuclear magnetic resonance (NMR) techniques to read induced spin states of atoms. A third is to polarize photons in an optical chamber. FinkelProducts has applied the second option in the QCPU, using NMR techniques to read specifically induced spin states in carbon, hydrogen, and other atoms (Josephson 1999, 301).

To manipulate carbon and hydrogen atoms, radio frequency (RF) energy is applied to each atom at its specific resonant frequency. This RF energy is applied to the atom while it is in a fixed magnetic field. Because the atom remains in a fixed position, the position can serve as its memory address. The nucleons of these atoms spin predictably while in this magnetic field. If an atom lines up with the direction of the magnetic field, it is considered to be in a "spin up" orientation. If it lines up in a direction opposite to the magnetic field, it is considered to be in a "spin down" orientation. Different spin states have different energy signatures for different atoms at different magnetic field magnitudes. These differences can be read by NMR sensors.

Discussion

As an example of the kinds of discussion material this type of report might contain, some information is also provided on the genesis of the QCPU. This kind of discussion would be useful for topics that deal with radically new technologies that vary significantly from traditional methods. Quantum computing is clearly such a topic.

Genesis of the QCPU
In 1998 George Yamaslute demonstrated that different RF frequencies cause predictably different spins for certain atoms. He also showed that the spin of these atoms can be altered by the application of different RF signals. These altered spin states then can be used symbolically to represent different values. The spin states of these atoms effectively store the values encoded by these RF signals. These values are then read by traditional NMR techniques, thereby creating a machine memory capability (Yamaslute 1998, 200–210).

Besides memory capabilities, manipulating the spin states of atoms can be used to perform various logic operations. Early experiments with molecules containing carbon and hydrogen atoms demonstrated such theoretical feasibility. The carbon and hydrogen atoms can be manipulated independently by varying RF signals applied at different resonant frequencies while these atoms are held in a fixed magnetic field. By making both atoms spin up, or spin down, or alternately spin up and spin down, the atoms can constitute a two-bit truth table. Adding additional types of atoms and using intermediate spin states substantially increases the range of logical operations. Such manipulation provides the quantum logic capabilities of the QCPU (Yamaslute 1998, 240).

Finally, the discussion should include a brief description of the QCPU device because the chip is the primary focus of this report.

The QCPU

The FinkelProducts QCPU is proprietary technology used today only in highly classified government projects. Although unclassified information is limited, some scientists believe that the QCPU is now providing the computational power for the genetic manipulation of bacteria. The specific goal is to create virulent strains of Naomi-B bacteria that are capable of eating through the armored titanium and steel hulls of enemy submarines at depths exceeding 10,000 feet (Mierson 2000, 50).

The QCPU assembly consists of the quantum molecular matrix (QMM), the magnetic field coil (MFC), the nuclear magnetic resonance sensor (NMR), and the RF assembly (RFA). The MFC provides the fixed magnetic field engulfing the QMM. The QMM provides atomic structures that have unique spin up and spin down characteristics. The RFA provides the RF energy needed to change the spin characteristics of each atom to reflect specific values (Bearkins 2000, 91). The energy is radiated by a phase directional array, but the exact design or method of control is proprietary. The NMR is the means of sensing or reading these altered spin states.

Conclusion

The conclusion section of a research report normally summarizes the report and may provide a recommendation. Any recommendation must be supported and justified by information in the discussion section. Given the nature of this sample report's theoretical discussion, a specific recommendation is not supported or justified, and one is not provided. In this example, the conclusion will only summarize the information provided on the QCPU.

Summary

The FinkelCHIP Quantum Central Processing Unit (QCPU) has made quantum computing a reality. By controlling and reading the spin states of selected atoms using applied RF and NMR techniques, the QCPU effectively uses quantum phenomena to store data and conduct logical operations on that data. Quantum computing, by exploiting the possibilities of data manipulation and storage at the atomic level, provides the power to accomplish parallel processing at far greater amplitudes than traditional approaches.

The actual design and implementation of the QCPU is not only proprietary but is also highly classified for national security purposes. Little information is available about the specifics of the QCPU's design implementation.

References and Appendix

Finally, include a list of the references used in the report. Always list all references that were actually cited in the report. As a courtesy to your reader, you can also list sources that you consulted but did not specifically use because these sources may have influenced your thinking. In this example, five sources were actually cited in the paper, and one source was consulted but not used.

Consequently, in the following section of this chapter, where the complete report has been assembled, you will note that two categories of references are listed: sources consulted and used and sources consulted and not used. For a more detailed discussion of documentation, see Chapter 13. By the way, this example also includes an appendix that contains the public relations (PR) release on the QCPU. If you need to include additional information that is not necessary for understanding the report (such as a manufacturer's specification sheet, PR release, schematics, or complex listings), it is best to do so in the appendix section of the report.

Putting It All Together	Here is the complete state-of-the-art research report, including visuals, references, and appendix.

State-of-the-Art Report on the FinkelCHIP Quantum Central Processing Unit

Introduction
Purpose
The purpose of this report is to provide a state-of-the-art investigation of the FinkelCHIP Quantum Central Processing Unit (QCPU), including a theoretical review of the premises of its operation.

Problem
Traditionally, computing power has been enhanced by increasing CPU speeds primarily through decreasing the size of conductors and solid-state devices used in chip fabrication. Decreasing size, however, has finite limitations, such as those associated with reducing the dielectric constants of the required materials. There has also been a move toward increasing the number of instructions executed for each clock cycle, especially with reduced instruction set (RISC) processors.

FinkelProducts International has transcended this traditional paradigm by developing the QCPU. The QCPU exploits and manipulates the quantum nuclear spin states of atoms. The QCPU is capable of performing very large numbers of advanced computational tasks simultaneously using the superimposition of multiple values encoded into the respective spin states of individual atoms. The resulting (effective) CPU speed is equal to or better than 500 GHz (Josephson 1999, 291). This effective speed makes the QCPU an ideal chip for processing-intensive tasks such as cryptographic factoring of large numbers, DNA sequencing in genetic research, and interactive, three-dimensional holographic imaging in advanced virtual reality systems.

Scope

The QCPU is built around proprietary information owned and protected by FinkelProducts International. Consequently, this report will be limited to the general theoretical approaches underlying the QCPU architecture; it will not investigate the actual methods used by the QCPU to implement these theoretical approaches.

Background

Theory

Quantum computing theory applies the knowledge of quantum physics to exploit subatomic phenomena of common elements to perform extremely complex computational tasks. When properly exploited, these phenomena provide a truly unprecedented ability for massively parallel processing (Ardvark 1999, 446–448). Several options exist to exploit quantum phenomena in this regard. One is to equate binary values to the ground and excited states of an atom. Another is to use traditional nuclear magnetic resonance (NMR) techniques to read induced spin states of atoms. A third is to polarize photons in an optical chamber. FinkelProducts has applied the second option in the QCPU, using NMR techniques to read specifically induced spin states in carbon, hydrogen, and other atoms (Josephson 1999, 301).

To manipulate carbon and hydrogen atoms, radio frequency (RF) energy is applied to each atom at its specific resonant frequency. This RF energy is applied to the atom while it is in a fixed magnetic field. Because the atom remains in a fixed position, the position can serve as its memory address. The nucleons of these atoms spin predictably while in this magnetic field. If an atom lines up with the direction of the magnetic field, it is considered to be in a "spin up" orientation. If it lines up in a direction opposite to the magnetic field, it is considered to be in a "spin down" orientation. Different spin states have different energy signatures for different atoms at different magnetic field magnitudes. These differences can be read by NMR sensors.

Discussion

Genesis of the QCPU

In 1998 George Yamaslute demonstrated that different RF frequencies cause predictably different spins for certain atoms. He showed that the spin of these atoms can be set by the application of different RF signals. These altered spin states then can be used symbolically to represent different values. The spin states of these atoms effectively store the values encoded by these RF signals. These values are then read by traditional NMR techniques, thereby creating a machine memory capability (Yamaslute 1998, 200–210).

Besides memory capabilities, manipulating the spin states of atoms can be used to perform various logic operations. Early experiments with molecules containing carbon and hydrogen atoms demonstrated such theoretical feasibility. The carbon and hydrogen atoms can be manipulated independently by varying RF signals applied at different resonant frequencies while these atoms are held in a fixed magnetic field. By making both atoms spin up, or spin down, or alternately spin up and spin down, the atoms can constitute a two-bit truth table (see Table 10.1). Adding additional types of atoms and using intermediate

Table 10.1 Spin state truth table

RF signal applied	Carbon	Hydrogen
Frequency #1	U	U
Frequency #2	U	D
Frequency #3	D	U
Frequency #4	D	D

spin states substantially increases the range of logical operations. Such manipulation provides the quantum logic capabilities of the QCPU (Yamaslute 1998, 240).

The QCPU

The FinkelProducts QCPU is proprietary technology used today only in highly classified government projects. Although unclassified information is limited, some scientists believe that the QCPU is now providing the computational power for the genetic manipulation of bacteria. The specific goal is to create virulent strains of Naomi-B bacteria that are capable of eating through the armored titanium and steel hulls of enemy submarines at depths exceeding 10,000 feet (Mierson 2000, 50).

The QCPU assembly consists of the quantum molecular matrix (QMM), the magnetic field coil (MFC), the nuclear magnetic resonance sensor (NMR), and the RF assembly (RFA). (See Figure 10.1.) The MFC provides the fixed magnetic field engulfing the QMM. The QMM provides atomic structures that have unique spin up and spin down characteristics. The RFA provides the RF energy needed to change the spin characteristics of each atom to reflect specific values (Bearkins 2000, 91). The energy is radiated by a phase directional array, but the exact design or method

Figure 10.1
QCPU assembly.
(*Source: George S. Yamaslute,* "*Magnets and Bits,*" Journal of Applied Magnetic Resonance 4:9 (September 1998), p. 260.)

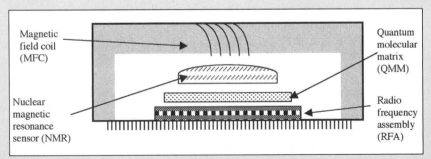

of control is highly proprietary. The NMR is the means of sensing or reading these altered spin states.

Summary

The FinkelCHIP Quantum Central Processing Unit (QCPU) has made quantum computing a reality. By controlling and reading the spin states of selected atoms using applied RF and NMR techniques, the QCPU effectively uses quantum phenomena to store data and conduct logical operations on that data. Quantum computing, by exploiting the possibilities of data manipulation and storage at the atomic level, provides the power to accomplish parallel processing at far greater amplitudes than traditional approaches.

The actual design and implementation of the QCPU is not only proprietary but is also highly classified for national security purposes. Little information is available about the specifics of the QCPU's design implementation.

References

Sources Consulted and Used

Ardvark, William J. "NMR and Beyond," *Quantum Computer Quarterly* 9:6 (August 1999), pp. 460–495.

Bearkins, Felinus S. "Cracking the Secrets of the FinkelCHIP," Internet: www.crackerbox.org, March 31, 2000.

Yamaslute, George S. "Magnets and Bits," *Journal of Applied Magnetic Resonance* 4:9 (September 1998), pp. 150–265.

Josephson, Albert E. "Technical Report 1999-35." FinkelProducts International, November 1999.

Mierson, Wilhelm F. "The Dreaded Threat of Naomi-B," *Journal of Truth in Science* 12:2 (February 2000), pp. 460–462.

Sources Consulted and Not Used

Baker, Joe-Bob. *A View of That Which We Cannot See.* New York: Acme Press, 1994.

Appendix

For Immediate Release

New York   August 31, 1999

FinkelProducts International has announced a stunning technological breakthrough, the FinkelCHIP Quantum Central Processor Unit. This device promises to revolutionize the way computers work by using individual atoms to store and process information.

The FinkelProducts Technical Team, which is known for its efforts to promote goodness and world peace, continues to develop and enhance this technology so that it will one day improve the quality of life for all who live on this earth.

-END-

Research Report Checklist	• Have I clearly stated the purpose of this report? • Have I introduced the topic with a brief overview of the problem or background?

- Have I discussed how I limited the report and my rationale for doing so?
- Have I provided adequate background for my reader to understand the report?
- Have I provided substantive, well-documented information in the report?
- Have I included necessary visuals and data?
- Have I summarized my research in the conclusion?
- Have I made a recommendation, and if so, is it supported by the discussion?
- Have I cited sources where necessary in the text?
- Have I listed the sources I cited?
- Have I included in an appendix any relevant material not necessary for understanding the paper?

11

Resumes and Interviews

The best way to get a job is to get your foot in the door by knowing someone important who is already inside. This approach is frequently called *networking*. When it is time to go out and get a job, or to move from one job to another, networking with the right people in the right places can be extremely effective. Unfortunately, for those of us who may not know these people, that approach will not work. If you fall into the latter category, you will have to start by developing a resume and cover letter. With these documents you can earn the opportunity to interview for a position and land the job.[1]

Generally speaking, resumes are not all that useful because they just do not work well. Few people who send out resumes actually get a job as a result. But that is *generally speaking*. For engineers and scientists, resumes can be effective—at least in landing an opportunity to interview. In particular, engineering and computer-related science positions tend to be based on requirements that demand specific skills such as fluency in Unix, background in homeostatic control systems, hands-on experience with electron microscopy, and the like. These are objective, verifiable qualities that lend themselves to resume presentations.

Effective resumes must contain specific kinds of information that relate to the position desired;

they are well written; and they are accompanied by an equally well-written cover letter. The resume and cover letter must be flawlessly edited. This chapter will show you how to produce an effective, professional resume and cover letter geared specifically to engineering and science positions. It will also give you some useful tips for what to expect in an interview.

What Is a Resume? *Resumes* are specialized proposals by which you offer your services to fill a position. Normally resumes include a short summary of experience and qualifications; however, engineering and science resumes need to do more than that. You need to match your specific skills clearly to an employer's needs.

Your resume should reduce an employer's uncertainty about the perceived risks involved in granting you an interview, making you an offer, and hiring you. Employers do not want to make the wrong decision and hire the wrong person. Those are expensive mistakes that employers try hard to avoid. Your resume should reassure them that they are not making a mistake with you.

Suppose a firm is looking for an entry-level electrical engineer to design microwave communication systems. You have the qualifications to do that job, but what kinds of information should you highlight to reduce an employer's uncertainty about hiring you? You would probably want to show the employer that

- You want to work for that particular employer.
- You want a position involving microwave communication systems.
- You have the needed technical background to do the job.
- You have the required written and oral communication skills.

- You have the necessary personal traits.
- You are available when the employer needs you.

The kind of resume we are talking about here contains more than just a short summary of your skills. You need to think of your resume as a sales document by which you convince a prospective employer that you are low risk—that you want to work in the position and that you will do a great job. To accomplish this, your resume should have the kinds of information called for in Outline 11.1.

Writing a resume is demanding work—but because you are dealing with your career, you should be motivated to put forth the effort. Fortunately, you already know a lot about the subject; but determining your strengths and developing the document are still challenging. It

Writing a Resume

Outline 11.1 Resume

Name	Your name, address, telephone, and e-mail address
Objective	The kind of position you want
Strengths	Your strongest skills and attributes
Education	Your formal education
• Degrees	
• Certifications	
• Honors	
• Relevant course areas	
Computer skills	Your computer literacy
• Operating systems	(in order of fluency)
• Languages	(in order of fluency)
• Applications	(in order of expertise)
• Platforms	(in order of expertise)
Experience	(in order of relevance to job objective)
Personal	What job-related personal information can you add?

is difficult to be objective when *you* are the subject. So take your time, and ask people you trust to read and respond to what you are writing.

Outline 11.1 provides a good approach for putting together a resume. Start with your name, address, and telephone number. As an example, we will use William E. McFinkel, a fictitious person with solid qualifications. It is a good idea to include an e-mail address if you have one, but only if you check it daily. Because McFinkel rarely checks his e-mail, his e-mail address should not be included. Doing so might make him seem unresponsive to an employer trying to e-mail him.

> William E. McFinkel
> 4141 E. 34th Street
> Selingford, Ohio 45121
> Home (513) 556-3432
> Work (513) 223-5699

Objective

Next, McFinkel needs a job objective that should clearly relate to the position for which he is applying. The objective should also be specific enough to reassure the prospective employer that McFinkel really knows what he wants. Consider this job objective:

> *Objective*
> Seeking a tough, challenging engineering position in radio frequency communication systems where I can help achieve corporate goals in a team environment.

What does this objective tell us about McFinkel? First, he knows what he wants. Employers like that, along with the fact that his objective matches the available position. Hiring someone is risky for any company; making a mistake can be costly and traumatic for all concerned. Because he knows what he wants to do, and because that seems to fit with the available position, McFinkel represents a good match for the employer. Consequently, it is likely that he will be happy working for the company, and that alone reduces the company's risk in hiring him.

Second, he is looking for a tough, challenging position. To an employer, he sounds like a real "doer"—exactly the kind of worker who will thrive in solving difficult problems for the company.

Third, he wants to help achieve corporate goals in a team environment. So he understands the nature of employment. He understands that companies exist primarily to make money. He also understands that he will be working in a team environment; technical problems are rarely solved by a single person.

Strengths

McFinkel needs to list the qualities that make him special and set him apart from others. Notice that the following strengths statement is written in phrases, not complete sentences. Phrases are acceptable here because they are economical, to the point, and easy to read.

> *Strengths*
> Special blend of technical, interpersonal, and communication skills and experience. Hardworking, creative, articulate. Able to work well with others to get the job done.

To an employer, this short statement effectively shows technical competence, the ability to work well with others, and effective communications as important qualities that distinguish this applicant.

Education

Frequently technical positions have specific degree requirements. This requirement may be a function of company policy or may be mandated by a client with whom the company is doing business. For engineering and science positions, always highlight your degree, not the school; employers are hiring your education, not your college or university.

Here is McFinkel's educational information:

Education
B.S. in Electrical Engineering with University Honors (June 2000)
Wright State University, Dayton, Ohio 45435

A.S. in Engineering Technology (June 1996)
Ocean Breezes Technical Institute, Ocean Breezes, North Carolina 27334

Relevant Course Areas
- Microwave Systems
- RF Systems
- Digital Computer Hardware
- Computer Networking Control Systems
- Linear Systems
- VLSI Design
- Pulse and Digital Circuits
- Software Engineering
- Data Structures
- Computer Graphics
- Comparative Languages
- Technical Writing and Speaking for Engineers

Notice that the degrees are shown first, followed by the schools, locations, and dates. Also notice that in the list of relevant coursework, the course areas most relevant to the objective are given first.

Computer Skills

Computers are essential tools in almost any workplace, and particularly in the technical and scientific workplace. Many engineering and science jobs are provided by small companies, which usually place a premium on multidimensional employees. So it is useful to note any computer skills you might have, even though you are not necessarily applying for a computer position. Always looking for any edge, McFinkel has done just that:

Computer Skills
Languages (descending order of fluency):
    C, C++, Ada, HTML, Pascal, Visual BASIC, COBOL, FORTRAN
Operating systems (descending order of fluency):
    Unix, SunOS, VAX/VMS, MS-DOS, Windows, MacOS
Hardware (descending order of expertise):
    Digital Alpha AXP; PC 486, Pentium and Pentium II/III; Macintosh PowerPC60X and G-3
Applications (descending order of expertise):
    Word, Illustrator, Photoshop, Excel, Powerpoint, Access, Navigator, Internet Explorer

Notice that McFinkel has some impressive computer credentials. (If you are going to invent a person, why not do it right?) However, understand that McFinkel is simply an example. *Never invent qualifications on a resume!* Lots of people lie on resumes, which is one of the reasons resumes are not all that effective; however, where technical qualifications are concerned, what you claim is easily verifiable. For example, if you say on your resume that you are fluent in C or Ada, all an employer has to do to verify your claim is ask you to implement a routine in C or Ada. That kind of thing happens in many interviews. If it happens to you and you cannot do what you said in your resume, the interview is over, and so is the job opportunity.

Experience

Nothing is quite as persuasive as a track record of success. Virtually any kind of successful prior employment can be valuable on a resume, but successful employment in a job that relates to the objective can provide a tremendous advantage. This is especially true if you can show that you exceeded the expectations of the job.

You do not need to list all your experience on a resume. On a job application form you probably will; but on your resume just list the experience that relates directly to the job objective. Space permitting, you can also list representative experience that does not relate directly to the objective.

McFinkel's resume lists his technical experience that relates to the job objective. However, his resume does not include his 14 jobs with various fast food restaurants or his work for the city cleaning up around the stadium after baseball games. That experience is simply not relevant.

Experience
4/94–Present: Engineering Technician, Computing Imaging Corporation, Dayton, Ohio. Responsible for installing and maintaining Alpha-based Digital 4000 computer systems.
• 1997 and 1998 Outstanding Employee of the Year
• 1995 and 1996 recipient of the IEEE Merit Scholarship
1/92–3/94: Repair Technician, Computer Service Center, Dayton, Ohio. Responsible for PC computer system software and hardware installation.
• 1993 Excellence in Maintenance Award

Notice that McFinkel's experience lists inclusive dates, job title, company, and location. It also summarizes the duties of the job and documents performance that exceeded expectations.

Personal

The final section of the resume is the catch-all personal category. You need to include job-related information here that may enhance your value to the company. For example, if you are not a U.S. citizen, that is OK—but if you are, by all means mention it. That means the company can bid you on certain contracts with the government or defense-related firms. Also describe your availability—the date you will be able to start work. It is also sometimes useful to indicate that you are willing to travel or relocate to meet company needs (if, of course, you are).

The final section of McFinkel's resume might look something like this:

Personal
U.S. citizen
Available June 2000
Willing to travel and relocate

Ten Tips for Creating a Good Resume

1. Be sure your resume is perfect both grammatically and stylistically. Pay excruciating attention to detail: there should be absolutely no errors. Employers look for communication skills, and, fair or not, they will key on any grammar or style error as evidence that you do not write well. Take advantage of grammar- and spell-checking software, but do not rely on it totally.

2. Tailor your resume for the position. With word processing software, this is a snap. Remember: even good resumes often do not work, and generic resumes *never* work. Be sure to create separate resumes for different types of jobs. A resume created by a computer scientist for a software engineering position will not work well for a network administrator position. Specifically, gear your objective to the job, and present the most relevant and supportive data first.

3. Always be truthful in your resume, but not self-effacing. In other words, do not say, "I have a serious heart condition, but so far I have been

stable with medication." Accentuate the positive about yourself, but do not go off the deep end. Do not write, "I am absolutely the most brilliant engineer, and I represent the quintessence of intellectual power and personal achievement." Use your common sense!

4. Limit your resume to two pages. If you can keep it to one page, that is even better. One way to keep it short is to avoid minor details that the employer already knows, or will not care about. For example, "references available on request" can be assumed. And non–job-related data, like the names or occupations of your family, or that you like to swim, dance, and work for world peace, are facts that no employer really cares about.

5. Do not include religion, politics, or fraternal organizations. This information can be risky and usually provides nothing relevant to the job objective. Additionally, it is usually a good idea to avoid mentioning hobbies, unless, of course, they are relevant to the job objective. For example, ham radio might be a hobby that is relevant to a microwave engineering job.

6. Do not use fancy paper or exotic folds, and do not paste a photo of yourself on the first page. Employers of engineers and scientists are usually very pragmatic and substance-oriented.

7. Avoid acronyms unless you are absolutely sure your reader knows them.

8. Never refer to yourself in the third person. It comes across as stilted and sounds weird.

9. Never put a date on a resume. If you do, it will seem outdated immediately.

10. Do not refer to salary requirements. Salary is a topic better suited to the job interview, which normally includes filling out a formal job application form before the actual interview begins.

Whenever possible, send your resume with a cover letter. However, simply including a note that says "Here is my resume; call me for an interview" will not be productive. In fact, this wastes one of the most powerful tools you have to reduce an employer's perceived risk of hiring you.

A good cover letter is every bit as persuasive as a good resume, and in some cases can even substitute for one. A bad cover letter often winds up in the trash, along with the resume attached to it.

An effective cover letter does four things:

Cover Letters

1. It demonstrates that you know about the company and really want to work for it.
2. It summarizes your key skills and experience that make you a desirable candidate.
3. It describes the personal traits that make you a desirable candidate.
4. It provides a conclusion that creates good will and invites a favorable response.

Look at McFinkel's cover letter, and notice how each paragraph has a particular function. First, start the letter with the name, address, and contact information used on the resume:

William E. McFinkel
4141 E. 34th Street
Selingford, Ohio 45121
Home (513) 556-3432
Work (513) 223-5699

Next, whenever possible, address the letter to a person, not an organization—and not to "Whom

It May Concern." Sometimes you will not have a choice; but if you can find out the name of the person running the office that has the job for which you are applying, use it. McFinkel's letter is addressed to the senior engineer who is supervising microwave systems development:

> Ms. Tracy Ann Burger
> Senior Engineer, Microwave Systems
> E-Wave Systems International
> 595 Miner Road
> Wutherford Heights, OH 44132

In the body of the letter, the first paragraph should demonstrate that you know about the company and really want to work for it. McFinkel has done some homework and shows that he is not just shotgunning this letter everywhere:

> While doing career research to locate companies that may have electrical and communications engineering opportunities, I read about E-Wave Systems International's government work with advanced satellite communications systems. In fact, the more I read about your company and its activities, the more convinced I became that I'd like to be part of the E-Wave team. To that end, I have attached my resume with the hope that you will give me the opportunity to discuss further how I might fit in and make a contribution at E-Wave Systems.

Notice how this first paragraph both demonstrates McFinkel's knowledge of the prospective employer and, as a bonus, also shows that McFinkel wants to be a team player. He has taken the time to learn about the company's work with the government on advanced satellite communications systems. And he clearly wants to explore how he might fit in with the E-Wave Systems team.

The next paragraph briefly summarizes McFinkel's skills and experience that make him a desirable candidate. Notice that he does not go overboard here because the details are available in the resume. This letter's goal is to get the employer to look at the resume:

> I am currently a senior in Electrical Engineering at Wright State University. My degree program, which I will complete with University Honors this spring, comprehensively blends electrical and radio frequency engineering concepts and principles. In addition, during the past few years, I have worked as an engineering technician for Microwave Products Corporation, with primary responsibility for maintaining waveguide calibration systems. I believe this has given me valuable experience with the same technology being used by your company in its microwave design activities.

The third paragraph reviews McFinkel's personal traits that make him a particularly desirable candidate:

> Personally, I am a strong, internally motivated self-starter with excellent analytical and organizational skills. I pride myself on being able to work well with others in a team environment and on having proven abilities in written and oral discourse. My technical communication skills are particularly strong.

One point worth noting: this letter is well written, and that alone supports McFinkel's claim that his communication skills are particularly strong. If McFinkel came across as inarticulate, that claim would be compromised, as would the entire resume package.

Finally, in the last paragraph, provide a conclusion that creates good will and invites a favorable response:

> Thank you for taking the time to review my resume. I realize how many letters and resumes you must receive, but I know I can be a valuable resource for your organization, and I would welcome the opportunity to interview at your convenience.

Notice that McFinkel didn't say, "I want a job—schedule me immediately for an interview!" You would be surprised how many people actually write something like this in their cover letters and then wonder why they get no responses.

Finding Jobs on the Internet

Increasingly, technical people are searching for jobs and posting their resumes electronically on various Internet job sites.[2] Some of these sites simply provide a public posting forum, whereas others actively match job skills to positions. Many sites are free to the job applicant, but others charge a fee—sometimes for each posting or search.

The efficacy of this on-line method varies considerably depending on job skills, regional demand, the particular site, and luck. Generally speaking, the information you provide on-line is similar to the information in a traditional resume; however, the format and specific content may vary. The best advice is to approach on-line job searches as a resource for finding employment but not as a replacement for more traditional job search methods.

Putting It All Together

Here is the complete resume package, including the cover letter and the formatted resume. Letter and resume formatting can vary, but it is generally best to use a traditional block or indented style that is normally available on word processors.

William E. McFinkel
4141 E. 34th Street
Selingford, Ohio 45121
Home (513) 556-3432 Work (513) 223-5699

Ms. Tracy Ann Burger
Senior Engineer, Microwave Systems
E-Wave Systems International
595 Miner Road
Wutherford Heights, OH 44132   March 14, 2000

Dear Ms. Burger:

While doing career research to locate companies that may have electrical and communications engineering
opportunities, I read about E-Wave Systems International's government work with advanced satellite
communications systems. In fact, the more I read about your company and its activities, the more
convinced I became that I'd like to be part of the E-Wave team. To that end, I have attached my resume
with the hope that you will give me the opportunity to discuss further how I might fit in and make a
contribution at E-Wave Systems.

I am currently a senior in Electrical Engineering at Wright State University. My degree program, which I
will complete with University Honors this spring, comprehensively blends electrical and radio frequency
engineering concepts and principles. In addition, during the past few years, I have worked as an
engineering technician for Microwave Products Corporation, with primary responsibility for maintaining
waveguide calibration systems. I believe this has given me valuable experience with the same technology
being used by your company in its microwave design activities.

Personally, I am a strong, internally motivated self-starter with excellent analytical and organizational
skills. I pride myself on being able to work well with others in a team environment and on having proven
abilities in written and oral discourse. My technical communication skills are particularly strong.

Thank you for taking the time to review my resume. I realize how many letters and resumes you must
receive, but I know I can be a valuable resource for your organization, and I would welcome the
opportunity to interview at your convenience.

Sincerely,

William E. McFinkel

Enclosure: Resume

William E. McFinkel
4141 E. 34th Street, Selingford, Ohio 45121
Home (513) 556-3432   Work (513) 223-5699

Objective
Seeking a tough, challenging engineering position in radio frequency communication systems where I can help achieve corporate goals in a team environment.

Strengths
Special blend of technical, interpersonal, and communication skills and experience. Hardworking, creative, articulate. Able to work well with others to get the job done.

Education
B.S. in Electrical Engineering with University Honors (June 2000)
Wright State University, Dayton, Ohio 45435

A.S. in Engineering Technology (June 1996)
Ocean Breezes Technical Institute, Ocean Breezes, North Carolina 27334

Relevant Coursework

Microwave Systems	RF Systems
Digital Computer Hardware	Computer Networking Control Systems
Linear Systems	VLSI Design
Pulse and Digital Circuits	Software Engineering
Data Structures	Computer Graphics
Comparative Languages	Technical Writing and Speaking for Engineers

Computer Skills
Languages (descending order of fluency):
  C, C++, Ada, HTML, Pascal, Visual BASIC, COBOL, FORTRAN
Operating systems (descending order of fluency):
  Unix, SunOS, VAX/VMS, MS-DOS, Windows, MacOS
Hardware (descending order of expertise):
  Digital Alpha AXP; PC 486, Pentium and Pentium II/III; Macintosh PowerPC 60X and G-3
Applications (descending order of expertise):
  Word, Illustrator, Photoshop, Excel, Powerpoint, Access, Navigator, Internet Explorer

Experience
4/94–Present
Engineering Technician, Computing Imaging Corporation, Dayton, Ohio. Responsible for installing and maintaining Alpha-based Digital 4000 computer systems.
• 1997 and 1998 Outstanding Employee of the Year
• 1995 and 1996 recipient of the IEEE Merit Scholarship

1/92–3/94
Repair Technician, Computer Service Center, Dayton, Ohio. Responsible for PC computer system software and hardware installation.
• 1993 Excellence in Maintenance Award

Personal
U.S. citizen
Available June 2000
Willing to travel and relocate

The purpose of a cover letter is to get an employer to read your resume. The purpose of a resume is to get an employer to interview you. The purpose of an interview is to get an employer to offer you a job. If you have the technical skills and personal characteristics an employer needs, and if you did a good job with your cover letter and resume, you will probably be invited to interview for the position.

It is important to realize that interviews are stressful and risky for both the interviewer and interviewee. Both parties have a lot to gain or lose depending on how things go. Interviews are decision points: the employer has to decide whether to give you a contract, and you have to decide whether to accept it.

Interviews come in all shapes and sizes, and they may have different purposes. Some are *screening interviews* designed to narrow the field to two or three top candidates; these are often conducted over the phone. They are usually followed by in-person *hiring interviews*. The final decision to hire is normally made based on the results of this interview. If you make the cut in the screening interview, you will be invited to the hiring interview.

Interviews can be as short as 15 to 30 minutes or can last several days. Some require only that you answer questions, whereas others require you to make a formal presentation about yourself and your work. Interviews can be formal, where you sit down in a room full of important-looking people and respond to one or more interviewers' questions. Or they can be very informal, such as a discussion at lunch or dinner.

Two things should occur in any interview: the employer should size you up to determine if you can fill the position; and you should evaluate the

Interviewing

employer to see if you want to work for that company. Note that when you report for an interview, you will usually be given a job application form to complete. Be sure to have complete contact information for each of your references with you, along with extra copies of your resume.

Here are some tips for successful interviewing:

1. Fully research the prospective employer before your interview. You should have done some of this research when you wrote your resume and cover letter. Now, however, you need more detailed information about the company, the kinds of work it does, and its track record of success. Also try to assess the quality of the people working there, how stable their employment is, what sort of salary and benefits package the employer offers, the cost of living in that area, and so on. For many employers, much of this information is readily available on the Internet or in a library. You can also request brochures and an annual report from the company before you interview. In some cases you can make discreet inquiries with those you trust who have experience with the company or who know people who work there.

2. Be prepared to answer not only technical questions but also personal ones. For example, if you were interviewing for a civil engineering position involving the design and construction of highways, an interviewer might ask you about cyclical loading of concrete and asphalt or fatigue and stress problems associated with aging bridges. But what if the interviewer asked you questions like these:

   • "What is your greatest strength and greatest weakness?"
   • "What are your long-term career goals?"
   • "Where do you see yourself in five years?"

- "Why do you want to work for this company?"
- "How well do you work with others?"
- "What do you know about this position?"

How would you respond to these questions? Think about that—these are probably the kinds of questions you will be asked first. Many new engineering or science graduates have no problem with the technical questions; but when asked the personal ones, they either turn nervous and incoherent, or they say something stupid.

3. Be prepared to discuss your salary requirements. You can find a variety of salary surveys on the Internet and in libraries. Identify the going rate for your particular skill level in the region where your employment will occur and decide, in advance, what your bottom line will be. On the job application form, you may be asked for your salary requirements. To keep from revealing your bottom line too early, just indicate on the form that your salary is negotiable.

4. Listen carefully in the interview. Take your time and respond to what is being asked. If you need to think about the question for a moment or two, say so. Do not blurt out a careless answer in an effort to respond quickly to a difficult question. Also, if you are not sure what is being asked, request clarification.

5. Look and act professional at all times. That does not mean you have to rent formal attire for the interview, but you should dress properly. For some companies, that may require a coat and tie, but find out in advance and do what is expected. Also act professional in your demeanor and the substance of your discussions. Be on time for your interview, do not make mindless comments, and do not speak disparagingly about former bosses and coworkers. Always be positive.

Cover Letter Checklist	• Have I demonstrated that I have researched the prospective employer and have a reasonable knowledge of the company?
	• Have I indicated that I really want to work for this company?
	• Have I summarized my skills clearly and succinctly?
	• Have I summarized my personal traits that enhance my qualifications for this position?
	• Have I demonstrated my desire to discuss further how I might contribute to achieving corporate goals?
	• Have I indicated my willingness to interview for the position?

Resume Checklist	• Have I put my name, mailing address, telephone number, and e-mail address (if appropriate) at the top of my resume?
	• Have I provided a clearly stated job objective that is *not* generic?
	• Have I briefly but effectively described my strengths?
	• Have I listed my higher education, including degrees, honors, and coursework?
	• Have I summarized my computer skills?
	• Have I listed my experience, putting first the most relevant to the job objective?
	• Have I included a personal section that provides, as appropriate, information about my citizenship, availability, and willingness to travel and relocate?

| **Interview Checklist** | • Have I researched the company, and do I understand the position? |
| | • Have I prepared myself to answer technical questions about what the company does? |

- Have I prepared myself to answer personal questions about my career goals and how working for this company in this position fits in with those goals?
- Have I consulted salary surveys, and do I know my minimum salary requirements?
- Have I dressed properly, and do I have an appropriate appearance for the interview?

---

1. For an extensive discussion of resumes, cover letters, and interviewing, see Richard Nelson Bolles, *What Color Is Your Parachute?* Berkeley: Ten Speed Press, 1999.

2. Internet career sites represent a dynamic resource that changes almost daily. However, some of the major career sites include the following: http://www.career builder.com; http://www.careermart.com; http://www. careermosaic.com; http://www.careersite.com; http:// www.collegegrad.com; http://www.intellimatch.com; http://www.monster.com; and http://www.4work.com.

Notes

12

Grammar and Style

This chapter is not intended as a grammar lesson; rather, the goal is to provide a practical guide to common grammar and style errors in technical reports—and specifically, to help you identify and fix these problems when they occur.

What Is Grammar, and Why Is It a Big Deal?

Grammar is nothing more than a large set of rules—commonly accepted standards for assembling words so that, together, they make sense and convey meaning. *Style* is the choice of words and the way we apply the rules of grammar in our writing. In traditional grade school curricula, grammar and style are the subject of much attention. What is important in technical writing, however, is not the ability to recite obscure grammatical rules; rather, it is being able to write correctly and effectively. In fact, grammar and style are important in technical writing for only two reasons:

1. Incorrect or improper grammar can change the meaning of what you are trying to say or, at least, make your meaning hard to decipher. That is fundamentally opposed to the goal of technical writing, which is precision in meaning.

2. Incorrect grammar says something about you and the quality of your thinking. Poor grammar in a technical report can communicate to your

reader that you are stupid or that you lack the required education or professional attention to detail.

In a technical document, you are judged to some extent based on the document's quality. By the time a reader is absorbing what you have written, you are not there to defend yourself, you are not available to explain what you really meant, and you have no opportunity to fix your errors.

Fortunately, most technical writers who make errors in grammar or style do so in relatively few areas. So this chapter will focus on the most frequent grammar and style mistakes in technical reports and will provide some straightforward solutions. Here are the most common problem areas:

- Comma splices
- Fused sentences
- Sentence fragments
- Misplaced modifiers
- Passive voice
- Verb agreement errors
- Pronoun agreement errors
- Pronoun reference errors
- Case errors
- Spelling errors

Comma Splices

Comma splices occur when we join one sentence with another sentence using a comma instead of a conjunction. This mistake is easy to make and easy to correct. Here is an example:

> Comma splice
>
> The circuit operates at DC, Ohm's Law applies.

Here we have two sentences: "The circuit operates at DC" and "Ohm's Law applies." The comma that follows "DC" is splicing these two sentences together, which is not something commas are supposed to do.

To fix the problem, you have two choices:

- Use a *semicolon* instead of that splicing comma:

  The circuit operates at DC; Ohm's Law applies.

- Or follow that splicing comma with a *conjunction*:

  The circuit operates at DC, and Ohm's Law applies.

Fused Sentences

A *fused sentence* is a comma splice without the comma. In other words, two sentences are fused without any mark of punctuation:

The workstation was not designed ergonomically it leaves much to be desired.

Again we have two sentences: "The workstation was not designed ergonomically" and "It leaves much to be desired." Note how they just run together at the indicated point of fusion.

The solution, as with a comma splice, is easy:

- Insert a *semicolon:*

  The workstation was not designed ergonomically; it leaves much to be desired.

- Or add a *comma* and a *conjunction:*

  The workstation was not designed ergonomically, and it leaves much to be desired.

- Or add a *semicolon,* an *adverb,* and a *comma:*

  The workstation was not designed ergonomically; consequently, it leaves much to be desired.

Sentence Fragments

For a sentence to be complete, it must contain a verb. *Fragments* usually occur when the writer substitutes something else for this verb or leaves the verb out altogether:

"Testing" is not a verb.

> Tensile *testing* the specimen carefully with high levels of precision.

This is a fragment because it does not contain a real verb. The word *testing,* derived from the verb *test,* does not act like a verb here; rather, it is a *gerund,* which is a verb used as a noun.

To fix this problem, you can convert the gerund *testing* into a real verb:

> The technician *was tensile testing* the specimen carefully with high levels of precision.

Or you can add a verb:

> Tensile testing the specimen carefully with high levels of precision *is* necessary.

Sentence fragments also occur when we put a subordinator before an otherwise perfectly good sentence:

> *Because* the transformer could not take the load.

The subordinator *because* makes this clause dependent on something else, but the something else is not there.

To fix the problem, you can remove the subordinator:

> ~~Because~~ The transformer could not take the load.

Or you can add the "something else":

> *Because* the transformer could not take the load, *the system quickly failed.*

Misplaced Modifier Errors

English relies on word order or placement for meaning. In effective technical writing, modifiers have to be close to the words they are supposed to modify. Sentences where modifiers are misplaced may be grammatically correct, but often they will not mean precisely what the writer intended. Here is an example:

> Ignorance of science is a phenomenon among students *that must be destroyed.*

Misplaced modifier

The writer is advocating destroying the ignorance of science among students. Because of the misplaced modifier, however, the writer is proposing that we destroy the students.

To fix this problem, move the modifier so that it relates more directly to what it is supposed to be modifying:

> Ignorance of science is a phenomenon *that must be destroyed* among students.

Passive Voice Problems

Passive and *active voice* refer to the movement of action through the sentence. In an *active* sentence, the subject comes first, then the verb, then the object of the verb's action. In a *passive* sentence, the object comes first, then the verb; the subject appears after the verb, if it shows up at all.

Consider the following active and passive sentences:

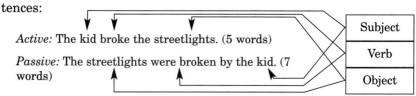

Active: The kid broke the streetlights. (5 words)

Passive: The streetlights were broken by the kid. (7 words)

Subject
Verb
Object

In the active sentence, the subject is the *kid.* The kid's action involved *breaking,* and the object of this action was *the streetlights.* The second sentence, the passive one, says exactly the same thing, but the object of the action comes first. The sentence is longer, and its construction is weaker. In addition, notice that the passive sentence's verb has an auxiliary verb—a form of the verb *to be.* Instead of just *broke,* the verb now consists of the verb *were* and the past participle *broken.* Passive constructions always pair a form of the verb *to be* with the past participle of a verb. That is a good way to spot passive voice, although such pairings are not always passive, and this method is not foolproof. The final test for passive voice is whether the object comes before the verb.

Passive constructions are weak, so why do people write in the passive voice? One reason is that it allows them to hide responsibility for their actions. Consider this modified passive sentence:

| Object |
| Verb |

Modified passive: The streetlights were broken.

What is missing? The subject, of course: whoever broke the streetlights. The sentence may be grammatically correct, but it leaves out an important piece of information. We no longer know who is responsible!

People sometimes use the passive voice to hide their culpability in something that is bad; then, in the same sentence, they switch to active voice to take credit for something that is good. Consider this compound sentence:

Your medical records were lost, but I found them.

Here the passive voice in the first clause hides who lost the medical records, but the writer has switched to the active voice in the second clause to take credit for finding them.

Active voice is preferred in technical writing because it is more direct, it is clearer, and it provides the most information with the fewest words. However, passive voice does have its place in technical writing. It can be useful when the subject of the sentence is unimportant or obvious or when the object of the sentence is the primary focus. Passive voice can also be a useful way of breaking the pattern of sentence structure to keep the reader from falling asleep. Finally, as mentioned, passive voice may be useful when we want to hide responsibility.

Verb Agreement Errors

Verbs must agree with their subjects in person and number. If the subject is in the first person, then the verb has to be a first-person verb. That is why "I *is* smart" is wrong, and "I *am* smart" is correct. Many verbs in the English language are not person sensitive, and for them, person does not matter.

In addition, if the subject is singular, its verb has to be singular. This requirement is the source of most verb agreement errors because, unlike agreement in person, some agreement-in-number errors do not sound wrong.

Here is an example:

The implant, along with its associated circuits, were inserted into the patient's chest cavity.	*Were* does not agree with *implant.*

The subject of this sentence is *implant,* which is singular. There is only one implant, but the verb *were* is plural. The verb does not agree with the subject, so a verb agreement error exists. This sentence may sound correct because the qualifier *along with its associated circuits* comes between the subject and the verb. Although *associated circuits* sounds plural, the verb must agree with the subject, not the qualifier.

You can fix this sentence either by making the subject plural or by making the verb singular. Changing the subject, *implant,* into its plural form might be grammatically correct, but technically it is not accurate—unless, of course, the surgeon is really putting two or more implants into the patient's chest cavity. Assuming that only one implant is being inserted, the only option is to make the verb singular, as follows:

The implant, along with its associated circuits, *was* inserted into the patient's chest cavity.

Pronoun Agreement Errors

Pronouns must agree with their *antecedents* (the words they replace) in person, number, and gender. This rule is somewhat similar to the subject–verb agreement rule. Consider the following sentence, in which the pronoun does not agree with its antecedent in number:

Plural pronoun with singular antecedent

Everyone in the lab must replace their radiation badges.

Believe it or not, *everyone* is singular (it comes from the two words *every* and *one*). The possessive pronoun *their* is plural, and trying to make it agree with *everyone* is a grammatical infraction. The solution is to make the antecedent plural or the pronoun singular. Consequently, either of these two sentences would be correct:

Singular: Everyone in the lab must replace *his* or *her* radiation badge.

Plural: All people in the lab must replace *their* radiation badges.

One other point worth mentioning here involves political correctness. Notice that plural pronouns in the English language are not gender-specific. That means you do not have to worry about what gender they are. It is generally safer to avoid specific gender references by using plural nouns and pronouns. "Their" is not gender-specific and, consequently, represents a safe approach—as long as the antecedent includes more than one person. If the antecedent refers only to one individual, use that individual's gender; if you do not know the person's gender, then use *his or her* or *her or his.*

Pronoun Reference Errors

The other common pronoun error in technical reports involves the use of a pronoun whose antecedent is unknown or unclear. In technical writing, pronouns must refer clearly and without question to a specific antecedent. Here is an example:

> The coolant leak impaired the CPU's heat dissipation, resulting in an erroneous reading at the most critical part of the process. *This* had a cascading effect on the system.

To what does *this* refer? It could refer to the coolant leak, to the erroneous reading, or to both. Such ambiguity can be problematic in technical writing, especially when you are using the pronoun *this.* In fact, a good rule is to always include a noun after the pronoun *this.*

You could fix this sentence as follows:

> The coolant leak impaired the CPU's heat dissipation, resulting in an erroneous reading at the most critical point in the process. *This coolant leak* had a cascading effect on the system.

Case Errors

Case errors involve putting a noun or pronoun in the wrong case. The three English cases are subjective, objective, and possessive. The *subjective case* (also called the *nominative case*) is what we put subjects in. The *objective case* is what we put objects in. The *possessive case* shows possession. Here is an example:

> I hit myself with my own tennis racket.

The first pronoun, *I,* is in the subjective case because it is the subject of the sentence. The second pronoun, *myself,* is in the objective case because it is receiving the action (*myself* is actually the reflexive form of *me*). The third pronoun, *my,* is in the possessive case because it shows possession of the tennis racket. In "I hit myself with *me* own tennis racket," the *me* represents a case error. (It would also sound ridiculous, unless, of course, I were Scotty in the old *Star Trek* series.)

Case errors can also get a little tricky in some instances:

> Needs to be possessive

The transmission microscope malfunctioning caused the experiment to be delayed.

In this example, the word *malfunctioning* is a gerund, or a verb used as a noun. A rule about gerunds says that their subjects are always in the possessive case. The subject of this gerund is *microscope,* which should be possessive in this sentence:

The transmission microscope's malfunctioning caused the experiment to be delayed.

Spelling Errors

Most people assume that an educated professional can spell words correctly. Spelling errors make your reader doubt the validity of your writing. So check and recheck your writing for spelling errors.

Consider how these errors would look in a technical paper:

> Wrong word

> Contraction for "it is"

It's important for a researcher two know the equipment and it's capabilities.

The *two know* obviously should be *to know,* and *it's* should be *its* (no apostrophe for the possessive pronoun that should be used here). Unfortunately, this type of error will usually be missed by your word processor's spell checker.

The correct sentence should read as follows:

> *It's* important for a researcher *to* know the equipment and *its* capabilities.

One additional spelling consideration: avoid starting sentences with a numeral:

> 9,192,631,770 Hz is the spectral line frequency of Cesium 133.

Correcting this kind of error can be difficult, but you have to get rid of that leading number because it is unattractive and can also be confusing in the middle of a document. The worst fix would involve spelling out the number:

> Nine billion, one hundred ninety-two million, six hundred thirty-one thousand, seven hundred and seventy Hertz is the spectral line frequency of Cesium 133.

An easier solution would be to simply insert an appropriate word before the number:

> *Exactly* 9,192,631,770 Hz is the spectral line frequency of Cesium 133.

Or simply invert the sentence:

> The spectral line frequency of Cesium 133 is 9,192,631,770 Hz.

When smaller numbers start a sentence, you can just spell them out:

> Thirty-two degrees Fahrenheit is the freezing point of water.

Finally, pay attention to which words you capitalize. Normally, in technical writing, some type of convention or style sheet will tell you what you should capitalize. Many style guides exist, along with lots of differences regarding the use of capital letters. Always follow the style guidelines for your organization or activity.

A good general approach is to avoid unnecessary capital letters. Consequently, if you are going to capitalize a word, have a specific reason for doing so. Here are some common reasons for capitalizing words:

- Capitalize names of specific persons, places, or things (proper nouns). These include the names of specific people (Albert Einstein); cities or street names (Dayton, Ohio, First Street); historical documents and religions (the Declaration of Independence, Catholicism); titles of papers, books, films, software packages, and trademarks (*Gone with the Wind,* Microsoft® Word); and periods of time or geographical regions (Cambrian Period, the West).
- Capitalize abbreviations or acronyms (ATM, USA, Ph.D.).
- Capitalize titles that precede a person's name, but not those that follow (Professor John Smith but John Smith, the professor).
- Capitalize the first word of every sentence and the pronoun *I.*

Exercise   How many errors can you find? All of the common errors discussed in this chapter, plus a few others, occur at least once in the following passage about the Finkasaurus Rex dinosaur. *Hint:* the following errors exist: 12 spelling, 1 case, 1 modification, 1 fragment, 5 pronoun, 3 verb, 2 comma splice, 4 passive voice, and 1 fused sentence. Additionally, a few minor punctuation errors also exist.

"The Finkasaurus Rex Dinosaur"

The Finkasaurus Rex dinosaur are a reptile known for their superier strength, intelligance, and good looks. Its name comes from the early greek words: *finkarium,* meaning "calculating"; *rexus,* meaning "king"; and *dinosaurs* meaning "terrible lizards" the Finkasaurus was the most terrible of all the terrible lizerds.

During the Triassic age of the Cretaceous period, the Finkasaurus Rex dinosaurs dominated the earth, they thundered across land, using their powerful leg muscles; they soared through the air, using small but efficiant airfoils formed by uniquely shaped dorsal fins; and they ruled the seas, often by stealth from beneath the waves, using there arms as diving planes. In fact, their submarineric depth and range was limited only by how long the dinosaur could hold their breath. 365 days a year, the dinosaurs feeding habits, along with those of its many friends.

The Finkasaurus Rex dinosaur have many distinctions, not the least of which was their extraordinary good looks. This is a clearly accepted concept.

The greatest of all saurischians, a rather unique shadow was cast by it's enormous skull and large trunk across the landscape, in fact, it is believed that when this great creature came out of hibernation on February 2, if their shadow could be seen by the smaller dinosaurs, six more weeks of winter would follow.

13
Documentation

Your technical writing teacher has just assigned a large research report that must have at least 10 sources, not including encyclopedias or Web sites. To save time and effort, you obtain everything you need from (where else?) an encyclopedia or Web site. Then, using that information, you whip up a complete research report. Next, through quick Web searches and a superficial skimming of newspapers, magazines, and books, you identify multiple sources that somehow relate, even tangentially, to the topics about which you have already written. These become the "sources" that you reference in your paper to keep the teacher happy.

The problem with this approach—besides leading to a paper with inferior content and throwing all standards of individual integrity and academic ethics into the dumpster—is that it compromises the role of documentation in technical writing. It ignores the need for complete, accurate source citations in the most serious form of business and scientific writing.

Technical writing often involves big business, lots of money, and a competitive, unforgiving environment. Proper documentation is an essential element of technical writing—an element that can have serious legal, ethical, and credibility implications for those who fall short of the mark. Documentation requirements are something that any technical writer must take seriously.

What Is Documentation? In its most general meaning, *documentation* refers to creating virtually anything recorded or "documented" on paper. All of the documents described in the second part of this book are forms of technical documentation. However, this chapter focuses on the kind of documentation that involves referencing sources. In this sense of the word, *documentation* gives formal credit to a person, organization, or publication for an idea or information that either is not original or is not common knowledge of the field. It represents an acknowledgment of your indebtedness to the source.

For example, if you need to quote the performance specifications for the FinkelTUBE from a FinkelProducts International technical report, you must document that technical report. Those specifications represent ideas that are not original—in other words, they are not *your* ideas. On the other hand, if you use Ohm's Law to show that 10 volts across a 1-ohm load produces 10 amps of current, you do not need to document Ohm's Law. Even though $I = E/R$ is not your idea, it is common knowledge in the field of electronics and, as such, does not need to be referenced.

Whether something is common knowledge of the field is often a judgment call. Normally, we think of something as being common knowledge when the average skilled person in the field should already know it. However, the best approach is to document any source when in doubt. Not only does this ensure that those who deserve credit receive it, it usually also enhances the credibility of your writing by adding some authority to the argument.

Your goal in providing documentation should be to give enough information about the sources you have used to enable your reader to find and consult those sources conveniently and independently. Sources typically include print media such as books, journals, periodicals, newspapers, conference proceedings, and dissertations; electronic media such as Web sites, File Transfer Protocol (FTP) sites, newsgroup postings, and CD-ROMs; and other material such as lectures and interviews.

Documentation Styles

Many style guides exist today for documenting sources. Here are a few: the *MLA Handbook for Writers of Research Papers;*[1] *The Chicago Manual of Style;*[2] the Government Printing Office *Style Manual;*[3] the *APA Publications Manual;*[4] the *American Chemical Society's Manual for Authors and Editors;*[5] the *American Institute of Physics Style Manual;*[6] and *The Council of Biology Editors Manual for Editors and Publishers.*[7] The best approach is to use the style guide specified by your employer, your teacher, or your field. If a style guide is not specified, you can use any consistent form of documentation that provides the necessary information.

When to Document Sources

As a technical writer, you should document sources for any or all of the following reasons.

To Meet Legal Requirements

Legally, when using copyrighted sources you are required to document these sources. Federal copyright statutes control the reproduction of original works, including books, music, drama, computer programs, databases, videos, sculptures, and virtually any other media. Although copyright laws do not specifically protect the ideas contained in these works, they do protect the expression of ideas by these works.

You may use copyrighted material, but only with permission of the copyright holder or, in some cases, without permission under the *Fair Use Doctrine*. Fair use provides for limited reproduction of copyrighted material without the permission of the owner for noncommercial, teaching, and research purposes. Fair use also requires that the original work be fully documented (referenced). If you do not provide this documentation, then it is not fair use—it is a violation of copyright law.[8]

To Meet Academic Standards

Academic standards require that you document any nonoriginal ideas, except those that represent common knowledge of the field. You must document not only direct and indirect quotations, but also paraphrases or any other discussions that specifically refer to or include original ideas that are not yours.

To Establish Credibility

You should also support your original assertions or conclusions, which are not based on common knowledge, when you can show complementing positions on the part of authoritative sources. The purpose here is to establish the credibility of your position. Avoid making unsupported assertions in technical documents. If your assertions are consistent with the ideas of a recognized authority or previous work, document (reference) that authority or work to establish your credibility.

How to Document Sources

Generally speaking, two different approaches exist for documenting: notational and parenthetical.

- *Notational documentation* places footnote or endnote superscript numbers in your paper at the point where you need to document a source. You would include the actual source citation either as a *footnote* at the bottom of the page or as an *endnote* at the conclusion of the report (or a sub-section of the report).

• *Parenthetical documentation* places a source citation in parentheses in your paper at the point where you need to document a source; you include a list of references at the end of the report (or major subsection). The citations are keyed to that list of references by either number or author's last name.

Most technical documents use parenthetical documentation because it is simple and effective. Parenthetical documentation also gives the reader more information at the citation.

Parenthetical documentation consists of two parts: the list of references at the end of the report and the parenthetical references within the text of the paper.

• The *list of references* (also called *list of sources, sources, references, notes, works cited,* or, frankly, whatever sounds reasonable) is essentially a bibliography that provides specific information (author, title, and publication) about the works used or considered by the writer. You can list the references alphabetically by the first significant word, which is usually the author's last name. Or you can sequentially number your sources. If you list your references alphabetically, you can usually reference the source with the author's last name. Numbering your sources will make things even easier because you can reference each source in the paper by its number in the list.

• The *parenthetical references* are inserted into the text of the paper as the source citations. In the parentheses, you first identify the source by either source number or author's last name, plus the date of publication, then add the specific pages being referenced (if applicable—some sources do not have page numbers). For example, consider the following paragraph:

> The FinkelTUBE is a beam power pentode normally used in Class C, radio frequency applications. However, as research by the Finkel-Laboratories so clearly demonstrates, the tube can be used successfully as a power amplifier in Class A audio applications as well. Its power, distortion, and signal-to-noise performance specifications are quite impressive in this regard. (3: 22–23) *or* (Yinburg 2000, 22–23)

This particular paragraph needs a reference to that FinkelLaboratories report. Either of the two sample citations will work. The first example uses a source number, whereas the second example uses the source's last name—in this case Dr. Robert W. Yinburg's last name because he is the chief scientist of FinkelLaboratories who actually wrote the referenced report.

As mentioned, the sources themselves are listed at the end of the report. This list of references might look something like the following:

List of References

1. Misra, Pradeep. "Order Recursive Gaussian Elimination," *IEEE Transactions on Aerospace and Electronic Systems,* Vol. AES-32 (January 1996), pp. 396–401.
2. Sudkamp, T. *Languages and Machines: An Introduction to the Theory of Computer Science,* 2nd ed. (New York: Addison-Wesley, 1996).
3. Yinburg, Robert W. "Laboratory Report: The FinkelTUBE in Power Audio Applications" (FinkelProducts International, September 2000).

The parenthetical documentation in our example was either (3: 22–23) or (Yinburg 2000, 22–23). Notice that both point to the same source in our list of references. Whichever parenthetical format you decide to use, be consistent throughout the entire report. Do not list a source by number in one section and by author's last name in another.

You may also want to add sources that you consulted but did not specifically use. The idea is not to pad your reference list, but to acknowledge sources that may have influenced your thinking even though you did not specifically cite them. These latter sources should be included in a separate list of references clearly indicating that they were consulted but not used. In other words, you might end up with a "List of Sources Consulted and Used" and a "List of Sources Consulted but Not Used"—or perhaps "Sources Cited" and "Sources Consulted."

Again, it is best to use the style guide specified by your boss, your teacher, or your field. However, if you do not have a specified format, the simple approach provided in this chapter should be adequate for documenting the most commonly used sources in technical writing.

Of course, several different types of sources exist, and each type is handled differently in the list of references. The following examples will show you how to handle the most common types of sources.

The following examples provide a general guide for documenting the most commonly used forms of print media.

Print Media Examples

Books

Include the author(s), title, edition, city of publication, publisher, and date of publication:

1. Sudkamp, T. *Languages and Machines: An Introduction to the Theory of Computer Science,* 2nd ed. New York: Addison-Wesley, 1996.

Journals

Include the author(s), article title, journal, volume (and number if necessary), date, and inclusive pages:

2. Brown, M.E., and J. J. Gallimore. "Visualizing of 3-D Structures During Computer-Aided Design." *International Journal of Human-Computer Interaction,* Vol. 7, 1995, pp. 37–56.

Conference Papers

Include the author(s), paper title, conference or transactions information, date, and inclusive pages:

3. Grandhi, R. V., and L. Wang. "High-Order Failure Probability Calculation Using Nonlinear Approximations." 37th SDM Conference, Salt Lake City, UT, April 1996, pp. 1292–1306.

Encyclopedias

Include article, encyclopedia, edition, and date:

4. "Capacitance Transducer." *Von Nostrand's Scientific Encyclopedia,* 5th ed., 1976.

Newspapers

Include author(s) if known, article, newspaper, date, section, and page(s):

5. "USAF Investigation Affects NASA." *Beavercreek News Current* (April 17, 1999), sec. A, p. 1.

Note: Because no author is listed, you would alphabetize this entry by the first significant word of the article's title, which is *USAF.*

Nonjournal Entries

Include author, article, publication, date, and inclusive pages:

6. Beale, Stephen. "New Plug-ins Add Hostlike Functions." *Macworld,* March 1999, p. 29.

Technical Reports

Include author, title, agency, number, and date:

7. Garber, Fred D. "Synthetic Aperture Radar Automatic Strategic Relocatable Target Identification System." Wright Laboratory, Wright-Patterson Air Force Base, Ohio 45433, WL-TR-93-1145, October 1993.

Dissertations and Theses

Include author, title, degree level, school, and date:

8. Rosa, Albert J. "Luminescent and Electrical Properties of Sodium Implanted Zinc Selenide." Ph.D. dissertation, University of Illinois at Urbana-Champaign, 1975.

The following examples provide a general guide for documenting electronic media. One problem you will run into, especially on the Internet, is that the publication or posting date will not be available, nor will page numbers. In that case, use the date on which the document was actually served to your computer—that is, the date you accessed the information. Indicate that this date is a "served" date, not a "publication" date. You can use a title or subtitle from the document in lieu of page numbers.

Electronic Media Examples

Internet: World Wide Web

Include author (if known), title, fully qualified URL (Web address), and date:

9.   "Stop! You Do Not Have to Write That Check to the IRS." Internet: http://www.irs.ustreas.gov/prod/cover.html, April 15, 1999.

Internet: Newsgroup

Include author (or topic), title, complete network address, and posting date:

10.   Hubble Space Telescope, "Daily Report #2351," USENET: sci.astro.hubble, April 15, 1999.

Internet: FTP Site

Include author, title, file name, complete network address, and date:

11.   Finkelstein, Leo, Jr. "College Recruiting Game" (ecsslots.exe), Internet: ftp.cs.wright.edu/~lfinkel, June 1998.

Computer Disk

Include file name, disk title, media type, version, series or ID number, and date:

12.   "PowerMac G3 ReadMe." *Power Macintosh G3—Minitower and Desktop Computers,* CD version 1.0, 691-2166-A, September 15, 1998.

Other Examples

Interview

Include interviewee, method, topic, affiliation, place, and date:

13.   Finkelstein, Leo, Jr. Personal Interview, Topic: "The Ethics of Using the FinkelKICK for Self-Defense." Office of the Dean, College of Engineering and Computer Science, Wright State University, Dayton, Ohio, June 11, 1999.

Lecture

Include lecturer, occasion, topic, location, and date:

14. Finkelstein, Leo, Jr. EGR 335 Class Lecture, Topic: "Documentation." College of Engineering and Computer Science, Wright State University, Dayton, Ohio, April 20, 1999.

- Have I used source citations throughout the text keyed to my list of references?
- Have I documented all uses of copyrighted material?
- Have I documented all nonoriginal ideas that are not common knowledge?
- Have I referenced authoritative sources that support assertions I have made that are not otherwise supported?
- Have I used the prescribed method and form of documentation (if applicable)?
- Have I been consistent in the method and form of documentation I have used?

Checklist for Documentation

---

1. *MLA Handbook for Writers of Research Papers,* 4th ed. New York: The Modern Language Association, 1995.

2. *The Chicago Manual of Style,* 14th ed. Chicago: University of Chicago Press, 1993.

3. *Style Manual,* rev. ed. Washington, D.C.: U.S. Government Printing Office, 1984.

4. *Publication Manual of the American Psychological Association,* 4th ed. Washington, D.C.: American Psychological Association, 1994.

5. *The ACS Style Guide: A Manual for Authors and Editors.* Washington, D.C.: American Chemical Society, 1986.

6. *AIP Style Manual,* 4th ed. New York: American Institute of Physics, 1990.

7. *Scientific Style and Format: The CBE Manual for Editors and Publishers,* 6th ed. New York: Cambridge University, 1994.

Notes

8. For more information, see the United States Copyright Office Web site. Internet: http://www.loc.gov/copyright/. In addition, a good resource for the new Digital Millennium Copyright Act (DMCA) can be found at http://www.utsystem.edu/ogc/intellectualproperty/dmcaisp.htm.

14

Visuals

Remember the last time you went to the dentist? As you sat in the waiting room, you probably picked up one of the magazines lying on the table. What did you look at? In all likelihood, you looked at the pictures. In fact, if you read anything, it was probably the captions under the pictures.

What Are Visuals?

Visuals are presentations of ideas that exploit one's sense of sight to communicate a large amount of information quickly and efficiently. You looked at the pictures first because of their information bandwidth; you could get more information from them than from reading the text, especially with the limited time available in the waiting room. You also looked at the pictures first because of their greater interest and higher credibility.

Technical writing deals with complex topics in precise ways. It is not surprising that one of the most important tools for a technical writer is *visuals*—things that we call *figures, diagrams, drawings, illustrations, graphs, charts, schematics, maps, photos,* and *tables.* Whether you are showing an exploded view of a mechanism or plotting the regression curve from an experiment, you will find that visuals are an absolutely essential element of any technical report.

General Guidelines for Using Visuals

- Include visuals in a technical paper only when you have a reason to do so. If you do not know why you are putting a visual into a paper, you probably do not need it.

- Be sure to reference a visual in the text discussion before its placement in the report. If the visual precedes its reference, the reader will wonder why it is there.
- Explain the significance of all visuals, supplying your interpretation of the data depicted.
- Be sure to number and title all visuals.
- Visuals must directly clarify or otherwise enhance the text discussion. You need to integrate them into your report, not just stick them somewhere. That means the labels and captions used in a visual should match the text discussion that refers to the visual. For example, if you are describing the negative terminal of a D-cell battery, do not call it the *negative terminal* in the text and the *cathode* in the visual.
- Be sure to document your visuals when they contain copyrighted information or represent nonoriginal ideas. Because visuals often get separated from the report, do not rely solely on notational or parenthetical documentation; include a source line with the visual itself.

Guidelines for Design of Visuals

Reproducibility

Design your visuals with the output process of your report in mind. If your report will be printed or duplicated in a single-color ink or toner, consider that fact when developing graphs and diagrams. Be especially wary of different colors that may look great on your video screen but could print with exactly the same shade of gray or even blend into the color of the paper on which the report is printed. A safe approach is to use pattern fills instead of colors when the report will be printed or duplicated in a single color.

Simplicity

Remember that the purpose of visuals is to clarify the information you are presenting. Some concepts (such as modeling the entire supply and distribution process of a large retail organization) are so complex that they do not lend themselves to visual presentation. Other concepts may need to be broken down into smaller components for effective presentation. For example, to show the operation of a laser printer, you would not include the charging electrode, electrostatic plate, scanning laser, toner reservoir, transfer roller, fusion roller, cleaning pad, and paper transport mechanism all in the same visual. Separate visuals for each part might be needed.

Accuracy

Ensure that any visual you use accurately portrays the information being presented. Do not exaggerate the data by manipulating scales or misrepresenting relative sizes. However, in certain documents, where it is clear that you are presenting a biased point of view, some visual enhancement of information is acceptable. For example, in a proposal, where you are selling your skills or your organization's ability to do a particular task, it is expected that you will enhance your strengths and play down your weaknesses. However, that does not mean it is acceptable to lie about your strengths or weaknesses.

Visuals generally fall into one of the following categories: diagrams, graphs, schematics, images, or tables. The following sections provide a brief discussion of each.

Types of Visuals

Diagrams

Diagrams are drawings that show the components of a mechanism, the steps of a process, or the relationship among parts of a system. Make diagrams only as complex as they need to be. Diagrams can provide normal, cutaway, or exploded views of a mechanism. Figure 14.1 provides a normal,

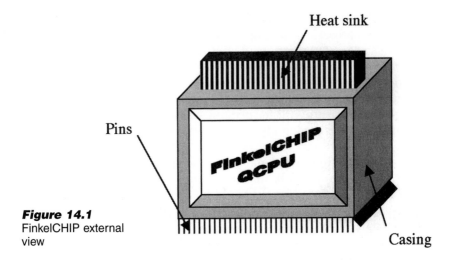

Heat sink

Pins

Casing

Figure 14.1
FinkelCHIP external
view

external view of the FinkelCHIP Quantum
Central Processing Unit discussed in Chapter 10.
This view would be useful for describing the exter-
nal physical attributes of the device.

Figure 14.2 provides a two-dimensional, cutaway view of the same device. This view would be
useful for describing the internal parts of the
FinkelCHIP and perhaps the process of its operation.

Figure 14.3 provides a three-dimensional,
exploded diagram of the same device. This view
might be useful for describing the physical
attributes of the internal structure, providing
assembly instructions, or describing the process
of its operation.

Graphs

Graphs are visual representations of relation-
ships among sets of numbers or of quantities and
proportions of mathematical values. Graphs are
great for presenting statistical information.
Generally, the three types of graphs you will

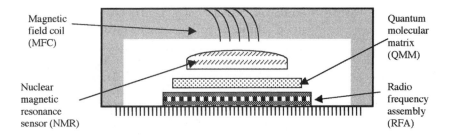

Figure 14.2
FinkelCHIP 2-D
cutaway

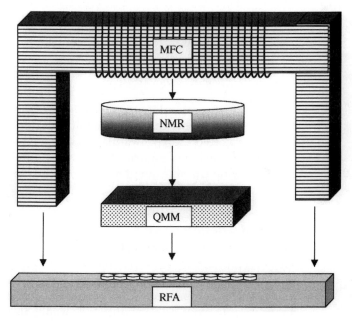

Figure 14.3
FinkelCHIP
exploded 3-D view

work with are line charts, bar/column charts, and pie charts.

Line Charts
Line charts effectively show trends in data. Normally, the vertical (*y*) axis is used to plot

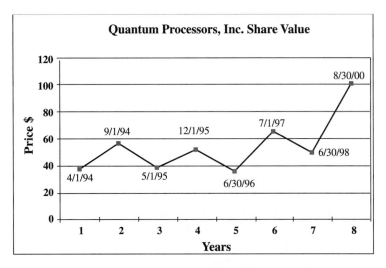

Figure 14.4
Line chart of critical point value trends (*Source:* FinkelProducts International; Technical Report 002345-S, August 2000, p. 21.)

dependent (variable) data points, and the horizontal (*x*) axis labels the independent variables. Each line plotted then shows the changing value of a specific variable. Figure 14.4 provides a line chart showing the six-year stock value trends for Quantum Processors, Inc.

Bar and Column Charts

Bar and *column charts* provide excellent tools for comparing discrete variables. Since each variable is a separate entity, a line chart would not be appropriate because no slope actually exists between data points. Normally, you would use the bar chart (Figure 14.5) when the value labels are too long to conveniently fit on the horizontal axis. You would use the column chart (Figure 14.6) when the value labels are short enough to fit well on the horizontal axis. Note that both of these figures provide discrete (and fictitious) data regarding the author's income sources.

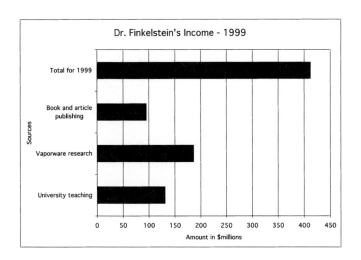

Figure 14.5
Bar chart of Finkelstein's income. (*Source:* "The Kind of Guy Your Daughter Should Marry," *Affluent American Magazine,* February 2000, p. 41.)

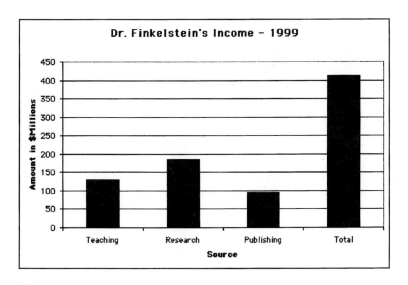

Figure 14.6
Column chart of Finkelstein's income. (*Source:* "The Kind of Guy Your Daughter Should Marry," *Affluent American Magazine,* February 2000, p. 41.)

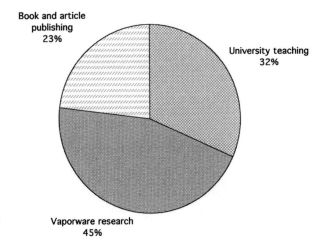

Book and article
publishing
23%

University teaching
32%

Vaporware research
45%

Figure 14.7
Pie chart of
Finkelstein's
income. (*Source:*
"The Kind of Guy
Your Daughter
Should Marry,"
*Affluent American
Magazine,* February
2000, p. 41.)

Pie Charts

Pie charts are useful for showing the relative proportions of a whole that each discrete data category represents (Figure 14.7). Pie charts, however, do not provide the same degree of visual
precision as line, bar, or column charts.

Adding Visual Interest to Line, Bar, Column, and Pie Charts

In some situations you might need to add visual
interest to a chart to get your reader to look at it.
A simple way to do that is to add a graphic or
photograph to the chart. Another way is to use
three-dimensional line, bar, column, and pie
charts. That extra dimension of depth will add
visual interest to the charts, but it can also make
the charts less precise. If your goal, however, is to
create dynamic impact, three-dimensional charts
might work well. Pie charts can also benefit from
such presentation, especially because they are
not used for precision anyway.

Dr. Finkelstein's Income for 1999

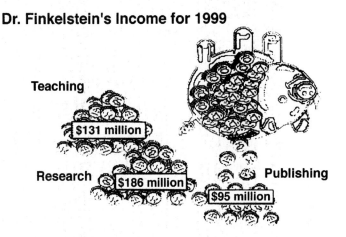

The most dramatic way to treat a technical chart is to turn it into a pictograph. Figure 14.8 is a pictographic chart of the author's (fictitious) income. This particular chart is analogous to the bar chart in Figure 14.5 and the column chart in Figure 14.6. Instead of bars or columns, this chart uses piles of coins as analogs for the values of each category. Obviously, this chart lacks the precision of either the bars or columns, but it is far more interesting.

Figure 14.8
Pictograph of Finkelstein's income. (*Source:* "The Kind of Guy Your Daughter Should Marry," *Affluent American Magazine,* February 2000, p. 41.)

Schematics

Schematics visually represent a system's structure or the procedures involved in a process. For example, the flowchart in Figure 14.9 provides a schematic of the process that some believe I used to write this book.

Schematics are also used frequently to provide circuit diagrams, such as the one of the Finkel-AMP's power amplifier circuit in Figure 14.10.

Figure 14.9
Schematic of book-writing process

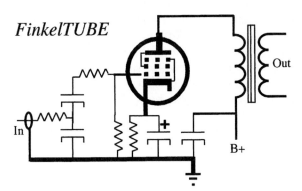

Figure 14.10
FinkelAMP power amplifier circuit

Images

Images are visual reproductions, such as photographs, that accurately reproduce the appearance of objects or events. They can add interest, credibility, and extreme visual detail to a technical report. Photographs are high-bandwidth tools that can give your reader lots of credible, nonverbal information quickly. For example, Figure

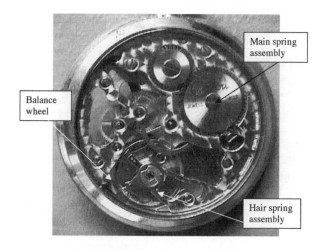

Balance
wheel

Main spring
assembly

Hair spring
assembly

Figure 14.11
Photograph of
watch movement

14.11 shows a macroscopic photo of an old watch movement. Such a photo might be used in an antique watch repair manual.

Modern digitizing technology has made the use of photography in technical reports relatively easy and inexpensive. When using photographs, however, be sure to double-check the intended output publishing process for the document. Photographs require grayscale or color information to be rendered properly in the final document; this may require special treatment and additional costs in the printing or duplication phase.

Tables

Tables are orderly arrangements of data and information in columns and rows. Tables are the most precise way to display an array of data. They also can be used for a wide range of technical writing applications, from simple listings to complex troubleshooting charts. Table 14.1 provides, in tabular form, the precise income data on which Figures 14.5 through 14.8 were based. Table 14.2 shows a troubleshooting chart for the FinkelKick (see Chapter 8).

Table 14.1   Dr. Finkelstein's income
for 1999

Source	Amount
Teaching	$131,002,130
Research	$186,000,456
Publications	$ 95,000,101

Source: "The Kind of Guy Your Daughter
Should Marry," *Affluent American Magazine,*
February 2000, p. 41.

Table 14.2   FinkelKick troubleshooting chart

FinkelKick troubleshooting chart

Problem	Causes and remedies
Wall did not break, and kicking foot was not injured.	You did not focus and accelerate properly. Foot did not hit the wall with adequate kinetic energy. • Focus better and kick harder!
Wall did not break, and kicking foot was destroyed.	You did not use the correct technique. • Practice your technique and then try it again with the other foot.
Wall broke, and kicking foot was destroyed.	You did not use the correct technique, but had adequate energy to break the wall anyway. • Declare victory and never try it again.

Table 14.3 shows a table that provides a simple listing of Finkelstein's purchases. Figure 14.12 is a pictographic representation of the same data.

Conclusion   Visuals are a key element of modern technical writing. They aid immeasurably in describing complex topics precisely by providing a large amount of information.

Table 14.3 Tabular listing of Finkelstein's purchases

Dr. Finkelstein: What does he buy?	
Toys	Friends
Food	Computers
Drink	Aphrodisiacs
Books	Videos
Travel	Facility maintenance
Utilities	Happiness

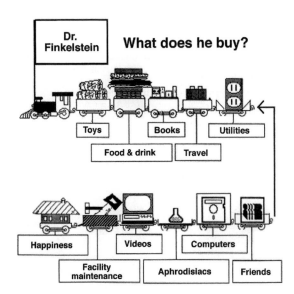

Figure 14.12
Pictographic of Finkelstein's purchases

Normally, we think of visuals as diagrams, graphs, schematics, images, and tables. Diagrams are drawings that show a mechanism or its components. Graphs, on the other hand, display and represent sets of numbers, quantities, and proportions. Schematics are visual representations of the structure of a mechanism or process, while images are accurate visual reproductions. Finally, tables display the orderly arrangement of data or other information in columns and rows.

Checklist for Visuals

- Have I selected the best visual for the kind of information I am presenting?
- Have I accurately displayed the information?
- Have I numbered and titled each visual?
- Have I documented the source if necessary using a source line with the visual?
- Have I integrated the visual with the text discussion that references it?
- Have I referred to the visual before its placement in the report?
- Have I used labels in the visual that match the terms used in the text?
- Have I considered the output publication process in my design of each visual?

15

Presentations and Briefings

If you believe your public speaking days are over simply because you have completed your last required speech course, you have a rude awakening ahead. The competitive world of science, technology, and business is built around technical presentations. Increasingly, companies are relying on their technical experts to present information to various audiences. The topics can be quite varied, from plans and programs, to goods and services, to advanced theories and concepts. The audiences can range from one's coworkers in a staff meeting to thousands of people in an international video teleconference.

Presentations and briefings are interpersonal performances in which concise technical information is provided to an attending audience. These performances, normally done live, are augmented by visuals and other media. Making effective technical presentations and briefings is not easy. Doing so requires a fertile imagination, a bit of courage, technical expertise in the subject, and the ability to communicate accurately and effectively.

When you are asked by your boss to make a technical presentation—and sooner or later, you will be—do not despair. Wasting energy on fear will not help you do a good presentation. What will help is the following:

What Are Presentations and Briefings?

- Identify the purpose for the presentation, and locate or develop the kind of information required.
- Find a simple way to organize these ideas clearly and effectively.
- Gear the presentation and its materials to the audience and purpose at hand.
- Practice delivering the presentation several times.

What makes a technical presentation effective is no big secret. Here are the key factors:

- Substantive ideas with precise, accurate information.
- Clear, coherent organization.
- Terminology and concepts appropriate for the audience at hand.
- Effective supporting media.
- Professional personal performance before the audience.

Substantive Ideas

The first consideration when giving a technical presentation is to have something worth presenting. In technical presentations the subject matter and purpose are usually well defined. You might be presenting a proposal on behalf of your company, the status of your project, the results of a laboratory study, or just information from a research project. In any case, the purpose and topic will dictate what substance needs to be included.

Clear, Coherent Organization

Once you figure out what you want to say, organize the material so you can present it effectively. Oral presentations differ from written ones in that spoken words are transient. If those reading your writing have trouble following what you are saying, they can reread it, think more about it, maybe ask someone else, and perhaps, in time, figure it out.

Those listening to an oral presentation, however, do not have that luxury. What happens if they get lost or cannot follow your line of thought? They may stop to think about it, in which case they will not be listening to what you are saying next. Or they may ignore what you said and try to keep up, in which case they missed your earlier point. They may just get frustrated and stop listening altogether, in which case, they are no longer part of your audience. Of course, people can always interrupt to say they do not understand; but in many situations, that is not likely. It requires an admission on their part, in front of their bosses and peers, that they do not understand what everyone else seems to understand. Normally, they will keep quiet.

That is why, especially for oral presentations, it is important to organize what you have to say in a way that is clear and obvious to your audience. Ideas—even difficult theoretical concepts—are much easier to follow and understand if they are organized logically and coherently.

Terminology and Concepts

Use words that are appropriate for your audience and type of presentation. Informal presentations can use informal language, whereas formal ones should be more "proper." Never use words that your audience will not understand. Readers can stop and look up something that they do not understand. But if your listeners do not accurately comprehend your words, you will lose their attention. Of course, they may also *misunderstand* what you are saying. This situation can be worse than when the audience does not understand, because now your listeners may believe they understand when, in fact, they do not.

Effective Delivery

You have figured out what to say, you have organized the material into a coherent and logical structure, and you have selected the proper language. Exactly how you deliver your presentation to the audience depends on the rhetorical situation. If you are speaking in an informal meeting, you might stand up and talk from your position at the table, or you might even stay seated. If you are making a formal presentation, you will probably do so from the focal point of the room, probably a podium or table up front. If you are giving a briefing, you will probably have charts—usually either transparencies with an overhead projector or computer-generated, presentation graphics running through a video projection system or large display terminal.

You may find yourself speaking unassisted to a small group, or you may use a sound system with thousands of people in the audience. Perhaps you will have a time limit. You may flip your own charts, or you may remotely signal a technician in a projection booth to change your charts. And, increasingly, you may speak through a teleconferencing camera to many people in geographically remote locations.

Obviously, you must adapt your presentation to whatever situation you are in. However, some general principles apply to any presentation. First, make sure you look and act professional. Also, if you have briefing charts, make sure your charts are professional in content and appearance. Some listeners may interpret substandard charts or inappropriate personal appearance as a lack of interest or capability on your part.

One final note regarding delivery of your presentation. It is normal to be nervous. To some extent, this nervousness is a positive thing: when controlled, it can give you an edge that will make your presentation more lively. However, nervousness, when not controlled, can distract your audience and degrade your presentation. It can cause you to stumble over your words, lose track of time, perspire on your notes, and speak too quickly.

So what do you do for nervousness? First, get to know your material backward and forward. Be familiar not only with what you plan to say, but with the theory and details behind what you are saying. Second, rehearse your presentation to the point where you are comfortable giving it. Practice may not make perfect, but it helps build your confidence and capability.

In the technical world, you will generally see three distinctly different types of speaking situations. Here they are, along with a few hints for dealing effectively with them.

Speaking Situations

Impromptu

Picture this: you are a systems administrator working for a large company. You have been invited to sit in the morning staff meeting. One of the VPs mentions hearing on the morning news something about a hacker breaking into a competitor's Web server and substituting a clown's picture for that of the CEO. Everyone laughs—except your company's CEO. With a deadly serious look on his face, he turns toward you and asks you to describe the security measures you are taking to counter such threats. Everyone turns, looks at you, and waits. You are on!

You have found yourself in an *impromptu situation,* where you have to talk intelligently on a complex topic with virtually no preparation time. Although such speaking would be highly unlikely in a formal situation, it does occur frequently in such informal settings as staff meetings. Clearly, this situation is risky. The wrong choice of words, topics, and arguments can be hazardous to your career. And if you say something really stupid, no one will remember that you did so in an impromptu mode—only that you said something stupid. So here are a few suggestions for handling impromptu situations.

First, if you think there is a chance you will be put on the spot, think through what you will say in advance. Clearly, as a systems administrator, you should have known about the hacker; and once you were invited to the staff meeting, you should have realized that the topic might come up.

Second, when unexpectedly put on the spot, do what forensic coaches teach their competitive speakers to do: buy some time to think about it. One way to do that is to divide the topic in some generic way. For example, everything has a past, present, and future. So, in this case, divide computer security into the past, present, and future—and talk, by way of introduction, about how it used to be in the days before networks and hackers. While buying time, you should be able to gather your thoughts for what you are going to say next. Of course, this assumes you know what you are talking about. If you are not sure, it is best not to make up facts in an attempt to bluff your way through the presentation. If you get caught, you will lose your credibility, which is difficult to get back.

Extemporaneous

Extemporaneous speaking is the preferred mode for technical presentation because the presentation is well prepared but not precisely scripted. If you give an extemporaneous presentation, you will follow an outline, but you will use your own words to discuss the material.

Extemporaneous speaking can be effective in technical situations, but only if you know the information and have practiced the presentation. You can use note cards for facts and figures and outlines for the presentation, but do not read from full pages of text. It is easy to get lost in a full page of text; and when reading it, you will often come across as stilted and insincere.

Manuscript

Manuscript presentations are totally prepared in advance. When you give one of these, all you do is read the script, with maybe some gestures and inflection added at appropriate places in the manuscript for emphasis. Try to avoid these kinds of presentations. They come across as insincere and overly formal, and when lengthy, they often generate boredom and despair among those stuck in the audience.

However, manuscripts do have a place in technical presentations, especially where high precision in detail and word selection is essential. For example, technical presentations that are going to be translated into several languages need careful word selection, especially where international diplomacy is involved. Also, if you are making a legal statement on behalf of the company, it is usually best to read, verbatim, what the legal staff has prepared.

One more point: *never* try to recite a presentation from memory. If you are distracted or have any kind of memory lapse, you will find yourself stuck in front of an audience with nowhere to go and nothing to say.

Knowing the purpose for a technical document is critical to how you write it. In the same way,

Speaking Purposes

knowing the purpose for a technical presentation is critical to what information you include and how you present it. Generally speaking, technical presentations are informative, demonstrative, or persuasive.

Informative

In an *informative presentation* your primary goal is to give the audience facts and other information. Informative presentations often take the form of a background briefing, where no decisions are required and no particular response is expected from the audience. A briefing on the capabilities of the FinkelCHIP would be an informative presentation. As you might imagine, these are relatively nonthreatening events that normally are considered low-risk.

Demonstrative

The primary goal of a *demonstrative presentation* is to show the audience how to do something or how something works. Teaching a computer class how to access and use a simulation on the Web would be such a presentation. These kinds of presentations often require audience interaction. They also tend to depend on having tools, equipment, and materials available during the presentation. Consequently, while relatively nonthreatening to the audience, they can be higher in risk because of your dependence on the equipment's working properly when needed. Test the equipment thoroughly *before* the presentation, understand what you are doing, and have a backup plan should something go wrong. Also, make sure you take all safety precautions, if warranted, so that you are not endangering anyone by your demonstration.

Persuasive

A *persuasive presentation* tries to convince the audience to make a particular decision or take some specific action. Often these presentations take the form of *decision briefings,* where you ask the boss to fund your project or approve an organizational change. Of all the technical briefings, these are potentially the highest risk because scarce resources, like dollars and people, are often involved. Your briefing may be part of a zero-sum game in which if you win, someone in the audience loses. The best advice is to know what you are talking about, have supporting facts and figures readily available, and keep your cool under fire.

Technical Briefings

Technical briefings are focused oral presentations that use visual aids normally referred to as *charts.* These charts can take the form of slides, transparencies, or computer-generated graphics. The briefing charts provide an outline of the presentation and, like technical documents, include words, illustrations, photographs, line graphs, and tables. They also add visual interest and transitions to the presentation.

General Guidelines

Use briefing charts to punctuate the presentation with short phrases and visuals, but do not attempt to provide a manuscript on the screen for the audience to read. Here are a few tips for producing briefing charts:

- Take full advantage of computer-generated presentation graphics whenever possible. Programs like Microsoft® Powerpoint are rapidly becoming the standard for business and technical presentations. Such programs are powerful software packages that provide many sophisticated capabilities, yet they are also generally simple to learn. You will find that producing truly professional presentations, using standard templates and color schemes, is easy and inexpensive with these software packages. You can also use these programs to print high-quality paper copies (often called *hard copies*), as well as backup transparencies, just in case the computer system goes down when it is your turn to speak.

- Make sure your charts will be readable in the room in which you will be speaking. Normally you will want to use at least an 18-point font. Avoid script and fancy fonts because they can be difficult to read. Standard templates that come with software packages like Powerpoint generally provide readable color schemes and font sizes.

- Pick your colors carefully. Lower-contrast combinations (like light blue on darker blue) may look fine on your video screen and on quality projection systems, but could visually fall apart with lower-quality systems. This problem can also occur when the ambient light in the room reduces the effective contrast of the projected image. Also avoid light-color fonts on dark backgrounds if you are not sure of the projection system and room. Under good conditions they can look dramatic; however, you will find that dark fonts on light backgrounds are readable in marginal situations when inverted combinations are not. Keep in mind that stronger colors like red can overpower some members of your audience, while weaker colors like yellow may fade out on them.

Like technical documents, technical briefings are straightforward and easy to organize. Typically a technical briefing contains the following charts:

Title chart	Summary chart
Overview chart	Concluding chart
Discussion charts	

Programs like Powerpoint provide separate layouts for title charts and the slide charts used in the body of the presentation. They also provide numerous color schemes and template designs, all of which can be customized with your own color preferences, artwork, and images. Pick a standard template, or develop your own specific color scheme and design. Then use that look consistently throughout the entire presentation.

Title Chart

The *title chart* leads off your briefing by telling your audience the topic of the presentation and your name and affiliation. In some cases you might want to add your e-mail address and telephone number. Figure 15.1 provides a sample title chart done in Powerpoint. This figure, and those that follow, use a custom template designed specifically for FinkelTECH (one of the author's many fictitious organizations).

Figure 15.1
Title chart

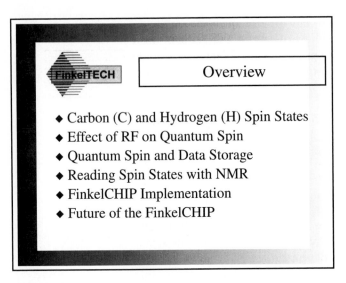

Figure 15.2
Overview chart

Overview Chart

The *overview chart* lists the main topics to be discussed in your briefing. It describes to your audience how you are organizing your presentation and what information will be included in it. Be sure to use short phrases (called *bullets*) on this chart, not complete sentences. Figure 15.2 provides a sample overview chart.

Discussion Charts

Discussion charts constitute the body of your presentation. Normally you will have one or more discussion charts for each topic listed on your overview chart. Figure 15.3 provides a sample discussion chart, which also includes a diagram to enhance the presentation and add visual interest. Note that discussion charts should also use short phrases (bullets) rather than complete sentences.

Summary Chart

The *summary chart* gives your audience a brief summary of the important points of your presen-

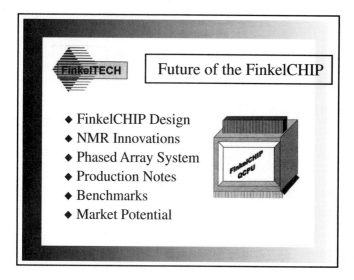

Figure 15.3
Discussion chart

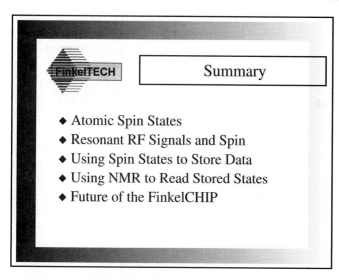

Figure 15.4
Summary chart

tation. In many cases you may be able to reuse the content of your overview chart as your summary chart. Figure 15.4 provides an example of a summary chart.

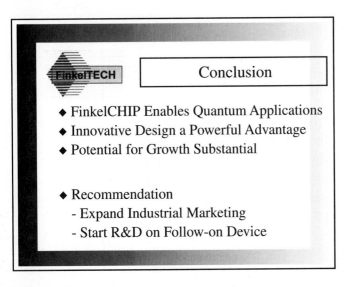

Conclusion

◆ FinkelCHIP Enables Quantum Applications
◆ Innovative Design a Powerful Advantage
◆ Potential for Growth Substantial

◆ Recommendation
   - Expand Industrial Marketing
   - Start R&D on Follow-on Device

Figure 15.5
Concluding chart

Concluding Chart

A concluding chart may or may not be required. In presentations where you are concluding something from your material and perhaps even making a recommendation (such as feasibility report presentations), you will need to add this chart. Figure 15.5 provides a sample concluding chart.

Controlling Complexity

A guiding rule for producing technical reports is to *use only the information that is necessary to get the job done.* This is even more important in technical briefings. Avoid adding anything to your briefing that does not have a specific purpose and serve a necessary function.

Visuals and Complexity

Your audience can absorb only so much information on the screen at one time. A service technician can effectively use, say, a large wiring diagram of a complex control system when that diagram is on his or her desk, and he or she has the time and need to trace through it. That same technician would find the same diagram useless, and even irritating, if you projected it onto a screen in a briefing. The detail would be difficult, if not impossible, to see or comprehend on the screen.

The same thing is true for complex tables or graphs of data. What if someone put the chart in Figure 15.6 on the screen and then talked about it for 20 minutes? The information is so complex, and the font size is so small, that not even an eagle sitting in the room could read it, much less a human being. And the human being and eagle would have about the same level of interest and understanding. If you have to drop the font size below 18 points to fit the information on the chart, then you are probably putting too much information on the chart.

The best approach is to decide if all that information is really necessary for the purpose of your briefing. If not, get rid of what is not needed. If

Figure 15.6
Unusable chart

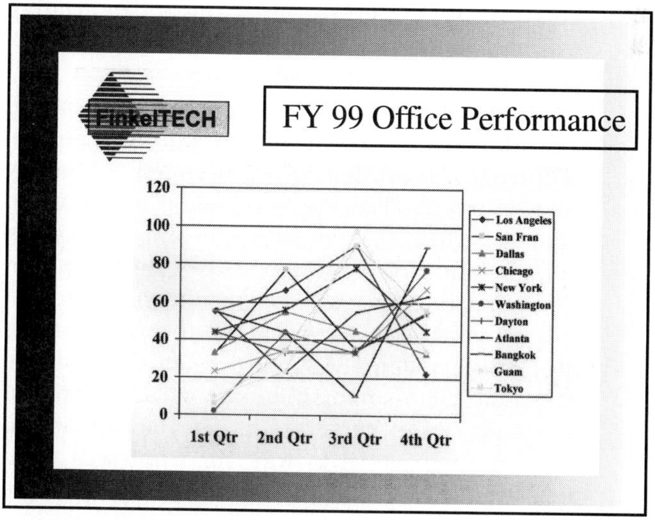

so, divide the data into multiple charts. More charts with less information are usually better than fewer charts with more.

Special Effects

Another way of adding complexity to briefings is by incorporating special effects. In presentation software packages like Powerpoint, it is a simple matter to add animation and sound to your charts. You can have text flying in and out, diagrams dissolving into one another, animated arrows directing the audience's attention, and a whole range of other visual effects. You can punctuate every visual effect with a pop, bang, or squawk. You can stir the emotions with big band music and soothe them with the sounds of the ocean.

The question here is not so much *what* you can do, but *why* you would want to use these effects. Special effects can be useful in small quantities—but be careful! Almost all of these effects get old quickly, even for the most patient audience. For example, using flying text or a humorous sound effect to highlight a point is fine, but using such effects throughout the entire briefing will become irritating and distracting. Too much of anything is usually bad!

There is one final consideration for special effects. Many of these effects are data intensive, requiring a great deal of processing overhead by the computer. That may not be a problem when you are sitting at home or in the office designing your briefing on your top-of-the-line computer. But what happens when you take your presentation on the road and the only available machine is an old 486? Your presentation may be unusable. The same thing can occur when you use high-resolution graphics in your presentation: you may have to wait for what will seem an eternity while the machine tries to load the next chart.

General Checklist

- Do I understand the occasion and purpose of the presentation?
- Am I presenting information that is substantive and relevant?
- Have I organized my information effectively?
- Am I using language and media that are appropriate to my audience?

Checklist for Technical Briefings

- Have I included a title chart with my name and organization?
- Have I included an overview chart listing the main topics of my briefing?
- Have I included at least one discussion chart for each topic on my overview?
- Have I included a summary chart (and conclusion chart if necessary)?
- Have I ensured that all my charts are readable when projected?
- Have I chosen an effective design scheme, and am I using it consistently?
- Have I included unnecessary complexity in any of my charts?
- Have I evaluated the briefing room and equipment beforehand (if possible)?

Checklist for Presentations

16
Electronic Publishing

In the old days, publishing was fairly straightforward. The author would write something, the copy editor would polish it, the layout editor would design it, the typographer would typeset it, and the printer would print it. The output of the process was a pile of paper designed and printed in a particular way and then bound and distributed to those who were included on a list to receive it. The publisher controlled almost everything about the publication: the fonts, the layout, the general appearance—and, to some extent, who got it.

Information technology has changed all that with *electronic publishing*. The electronic distribution of printed material and the elaborate networking of information are part of the general move toward a "wired" society. Traditional printed materials (such as periodicals) are also being distributed increasingly in electronic form. In effect, paper and ink, as the primary communication media, are being augmented, and to some extent displaced, by hypertext files and digitized images. Technical writing is part of this revolution in electronic publishing; technical writers today can no longer limit themselves to writing and preparing documents solely for traditional printing and duplication.

This chapter provides a brief orientation and introduction to electronic publishing, along with a few general considerations.[1] The information and guidelines presented throughout this book are just as valid for electronic publishing as for traditional publishing. However, there are significant differences in the way one designs, organizes, and distributes material when it is being published electronically. These differences are the primary focus of this chapter.

What Is Electronic Publishing?

Electronic publishing is the process of distributing information, sometimes on a wide scale, through computerized storage media and networks. At its most basic level, it involves encoding and electronically distributing printed documents using word processing files (such as those for Microsoft® Word) or specialized document description files (like those that are viewed with Adobe® Acrobat). Normally these files are designed to be loaded into a computer and either viewed or printed.

In many cases, such documents are distributed over the Internet using *file transfer protocol (FTP)*. In other cases, these documents take the form of either e-mail messages or encoded attachments to e-mail messages. These methods provide a fast, reliable means of moving complete files from one place to another; in fact, it is now possible to publish and distribute electronic documents more quickly than a traditional printing press can produce a single page. You can publish an electronic document, then, by placing it on an FTP server, or by distributing it through an e-mail list server, so that potentially large numbers of people can access the file and use it.

Hypertext Documents

A more sophisticated form of electronic publishing encodes and electronically distributes information in a totally different fashion from what would normally constitute a printed document. Most often, this type of publishing uses hypertext documents encoded with standardized, cross-platform markup languages. *Hypertext* is a method of representing words, images, sounds, and other on-line resources in a way that machines can read and that allows people using these machines to move directly from one text component to another by pointing and clicking (usually with a mouse). Web browsers, such as Microsoft® Internet Explorer and Netscape® Navigator, interpret these hypertext languages. In doing so, they allow Web surfers, using *hyperlinks,* to move rapidly from site to site, page to page, and within a page, accessing text, graphics, tables, and diagrams, as well as playing sounds and video.

Several markup languages exist. For example, the *Standard Generalized Mark-up Language* (SGML) is an international professional standard for information processing and office systems. The more common *HyperText Mark-up Language (HTML)* is the standard language of the World Wide Web and is the language that all Web browsers use. Other markup languages also exist, including the *Virtual Reality Markup Language (VRML),* which provides the capability to transmit and represent moving images of three-dimensional (3-D) objects. For example, an HTML document might allow a Web user to browse through a store's on-line catalog listings, whereas a VRML document could allow that same user to browse through a 3-D representation of the store!

In addition to the Internet and other computer networks, hypertext documents can be distributed in many ways. Optical storage media are commonly used to distribute electronic files. For example, a *compact disc read-only memory (CD-ROM)* can typically store up to 650 megabytes of data, while a *digital versatile disc (DVD)* can store up to 17 gigabytes of data. Given the capacities of these devices, it is a simple matter to publish an entire set of books on a single disk, including multimedia graphics, sounds, and full-motion video. DVDs not only have the capacity to store all those books, they can also hold full-length, wide-screen motion pictures, with sound tracks in multiple languages.

Many technical documents are now distributed in fully searchable, hypertext form on CD-ROM. For example, Microsoft® provides professional-level technical support through a set of CD-ROMs called *Technet.*[2] These CD-ROMs contain an extensive database of instructions, technical notes, and seminars, as well as a current *knowledge base* of data developed from troubleshooting experience. The CD-ROMs also contain not only descriptions of problems and solutions but, in many cases, the actual software patches and drivers to implement those solutions.

In addition, user-level technical support is frequently provided by computer and software manufacturers in hypertext files embedded within application programs. Normally called *on-line help* or *balloon help,* these technical support files often provide a complete on-line user's manual. Unlike printed manuals, on-line help manuals are readily available and often can be searched by phrase, keyword, or full text. And they can be changed easily and inexpensively with every update of the software package.

Producing Electronic Documents

Producing electronic documents is not difficult, but it requires some understanding of computerized media and how these media work. Hypertext documents contain the same information as traditional documents but use a different philosophy of organization. Much of this difference relates to the nature of electronic media and the capabilities a user requires to search for and link to desired information.

Normally, hypertext documents are similar to those described in this book but are organized into parallel, interrelated structures. This approach is unlike the traditional way of doing things, in which documents are organized sequentially, reflecting the physical assembly of their pages.

Converting Traditional Documents into Electronic Ones

Many electronic writers initially create their technical reports in the traditional format, then convert them into hypertext by adding HTML tags. Figure 16.1 shows the material you are now reading converted into HTML by Microsoft® Word. You will normally find it easier and quicker to use your word processor to convert your standard document into hypertext, especially if you are not an HTML programmer. You may also want to divide the document and save it as several smaller word processing files first, then convert these files into counterpart HTML files that can be interlinked. That would reduce loading times, as will be discussed later. Or you can just use a single, larger HTML file and define internal hyperlinks within that document. In any case, you can use any Web authoring or editing tool to format and reorganize these converted materials and to establish the hyperlinks among or within them.

If you convert your technical document into HTML, you will find that the resulting document does not take advantage of the powerful linking capabilities of hypertext. You will have only documents that can be viewed as single pages by a Web browser. If you do not divide the original word processing document into subsections before converting it to HTML, you will wind up

```
<HTML>
<HEAD>
<META HTTP-EQUIV="Content-Type" CONTENT="text/html; charset=iso-8859-
1">
<META NAME="Generator" CONTENT="Microsoft Word 98">
<TITLE>Producing Electronic Documents</TITLE>
<META NAME="Template" CONTENT="Macintosh HD:Microsoft Office
98:Templates:Web Pages:Blank Web Page">
</HEAD>
<BODY LINK="#0000ff" VLINK="#800080">

<B><FONT FACE="NewCenSb"><P>Producing Electronic Documents</P>
</B></FONT><P>Producing electronic documents is not difficult, but it
requires some understanding of computerized media and how these media
work. Hypertext documents contain the same information as traditional
documents but use a different philosophy of organization. Much of this
difference relates to the nature of electronic media and the
capabilities a user requires to search for and link to desired
information.</P>
<P>Normally, hypertext documents are similar to those described in this
book but are organized into parallel, interrelated structures. This
approach is unlike the traditional way of doing things, in which
documents are organized sequentially reflecting the physical assembly
of their pages.</P>
<B><FONT FACE=" NewCenSb"><P>Converting Traditional Documents into
Electronic Ones</P>
</B></FONT><P ALIGN="JUSTIFY">Many electronic writers initially create
their technical reports in the traditional format, then convert them
into hypertext by adding HTML tags. Figure 16.1 shows the material you
are now reading converted into HTML by Microsoft&reg; Word. You will
normally find it easier and quicker to use your word processor to
convert your standard document into hypertext, especially if you are
not an HTML programmer. </P></BODY>
</HTML>
```

Figure 16.1
Sample HTML
document

with one larger hypertext file that can be viewed only as a single page, but through which you can scroll up or down. Of course, you can still add all the necessary links whether you are working with a single file or multiple files.

Your new hypertext page probably will not look like the original document when you view it with a browser. Browsers handle formatting differently, and formatting capabilities are much more limited than with traditional word processing and desktop publishing. To some extent, this really does not matter because you will probably need to rework your document anyway. Converting your word processing documents into hypertext files represents just the start of the electronic publishing process.

Next, you will need to set up hyperlinks in your document to provide parallel, relational access to each section. The goal is to set things up so that the reader can quickly and effortlessly move between parts of your document. The easier you make it to locate and go to key parts of your document (or link to other documents), the more effective your design will be. But do not go off the deep end creating hyperlinks! They can quickly add more complexity and confusion than they are worth.

How Web Browsers Work

When creating hypertext documents, you must understand how a Web browser displays information. When you tell your browser to display something, you normally do so by entering a Web address, called a *universal resource locator* (URL), or a local HTML file name and path on your hard drive. Your computer must load the file that is going to be displayed, along with its associated resources (graphic images and so on), into your computer's memory. Only then can the file be displayed completely. In other words, if resources such as photographs, diagrams, or charts are part of the file to be displayed, the browser must read all of these resources into memory, as well as the HTML file itself, before the complete page can be viewed. (Many browsers allow you to turn off images and load text files only. In that case, the associated images will not be loaded or displayed.)

This requirement for browsers to load files is important when you organize your material. Remember that your reader's computer usually must load everything that is going to be displayed on a page before it can display the page. If your reader is accessing the information directly from a CD-ROM or an application help file or over a high-speed data network connection, this is no big problem. But what if your reader is accessing the file over a modem? Or what if the network is busy? By including excessive graphics in a single page, you unwittingly may make your page virtually unloadable. Think about that when building large hypertext documents or when including large graphics in your documents. Normally it is much better to include links to several smaller documents than to put everything in a single file.

Another fact to keep in mind: Once you convert a technical document into hypertext, you lose some degree of control over its formatting. The receiving computer's browser will control how your document appears. Even though some standardization exists with default font sizes and colors and the way HTML tags are handled, the displayed results still vary between browsers and video screens. That is why, when designing hypertext documents, it is a good idea to view them on several different browsers and screens first to ensure that the results are acceptable. Some browsers allow users to configure their own formatting. If your user has weird tastes in fonts and colors, your document might wind up looking quite different than what you had in mind!

Guidelines for Organizing Hypertext Documents

Traditionally, as mentioned earlier, technical documents are organized sequentially, from the first page to the last page, because that is how pages go together. As readers move through a technical report, they pass in an orderly way from one section to the next section, and within sections from one subsection to the next subsection.

This format works well with mystery and suspense novels. In technical reports, however, there is no mystery or suspense (at least, there shouldn't be!). So we try to mitigate the limitations of sequential access by adding devices like indexes, dividers, and tabs—all of which are designed to give the reader direct, parallel access to sections or subsections. These devices allow the reader to more readily locate the desired information. In hypertext reports, we do not need to include such devices because hyperlinks and search capabilities can take readers quickly and precisely to the information they are seeking.

The trick to electronic publishing, then, is to properly organize the material to expose the maximum number of hyperlinks, at any given time and place, to the most essential information in the report. This technique is particularly important at the top level of an electronic report. All the important links should be visible, or obviously available, when the top-level page is displayed. In other words, you should see the links when the page is loaded, or you should have a clear indication that other links exist and that you can scroll to them.

To understand how to reorganize a traditional report into an electronic, hypertext report, first go back to Chapter 9 and look at the FinkelTUBE laboratory report. The traditional document's organization is shown schematically in Figure 16.2. Notice how the traditional arrangement of the document is sequential, reflecting the successive way the pages are physically assembled. The flow starts at the beginning with the introductory statement of purpose, problem, and scope, and then proceeds through the background discussions on theory and prior research. Test and

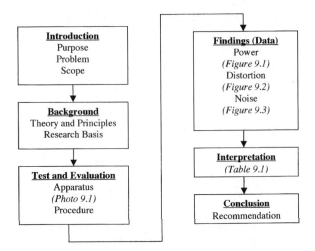

Figure 16.2
Traditional
organization

evaluation methods are discussed, including the apparatus and procedure, followed by the findings for power, distortion, and noise. The report ends with an interpretation of the results, then a conclusion and recommendation. We could mitigate this linearity with several techniques and devices, including a table of contents, an index, and dividers with tabs to help the reader locate the individual sections of the document. But the document would still be assembled sequentially.

Consider how we might reorganize this material when putting it into hypertext form. We are no longer constrained by the sequential assembly of physical pages, so we can think about putting important information in parallel form. Figure 16.3 provides a schematic of how this kind of parallel arrangement might work. Instead of sections of the report dictated by the physical requirements of page assembly, we now have screens that can be interconnected through hyperlinks.

Again, the goal is to get all the major links on the top-level screen. Figure 16.3 shows, in a simplified way, the parallel, hyperlink structure for

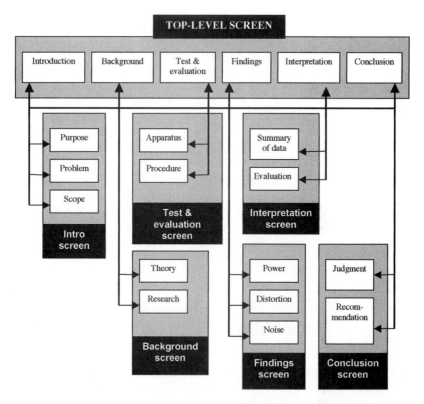

Figure 16.3
Hypertext
organization

the first two levels: the top-level screen and the screens immediately below this level. In reality, this electronic document would contain several more screen levels, with many more hyperlinks pointing back and forth among all the significant sections—and perhaps directly to information embedded within sections and subsections.

The design of the actual page can be quite varied and, to a great extent, is limited only by your imagination, creativity, programming capability, and common sense. Modern Web authoring software, like Microsoft® FrontPage, typically comes with a variety of templates and color schemes. You can also surf the Net to get ideas for layout and design. However, the best rule for electronic reports is to keep them simple.

Figure 16.4 provides a top-level screen for this electronic laboratory report on the FinkelTUBE; the figure shows what the page might look like when viewed with Netscape® Navigator. The six buttons at the bottom of the page provide primary, parallel links to the major sections of the report. Each of those sections would have its own page, which, in turn, would contain the information or provide hyperlinks to additional pages containing information. Links could also be set up to go directly to specific pieces of information. Of course, all pages would contain links to other major pages, as well as links back to the top-level page.

Figure 16.4
Top-level page

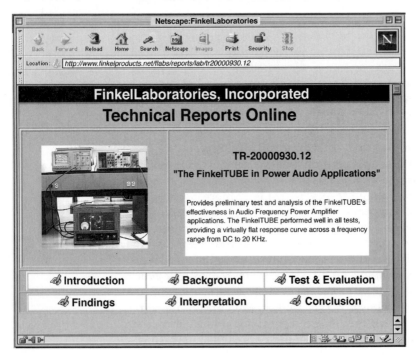

As discussed in Chapter 13, copyright laws provide federal and international protection for the expression of ideas, but not for the ideas themselves. Before electronic publishing and the Internet, the expression of ideas basically took the form of images printed on a page, whether those images were text, photographs, or illustrations. In other words, the ideas were encoded into the traditional print media and were protected in that form. In many cases it was impossible to separate an idea from the its expression (as with, say, a photograph)—and publishers exercised substantial control over what the expression looked like, what form it took, and where it was distributed.

In effect, the limitations of the medium provided reasonably good protection for the ideas it carried. You could copy or duplicate a printed page, an illustration, or a photograph—but you had to have an original or good-quality copy to do so. After one or two generations of duplication, the copy would become so degraded that it would not be worth having. And, of course, copying usually is not free. These inherent limitations prevented the widespread pirating of copyrighted material that we are seeing today.

Today, when an electronic document is placed on a Web server, it may well be accessible to anyone from anywhere in the world. Just viewing it requires that it be downloaded and copied to the viewing computer's memory—and probably, as well, to the browser's cache on the hard drive of that computer. Anyone viewing a Web document does not need to make a local copy because, given the nature of the technology and the way Web browsers work, that copy already exists. The person need only specifically name and save it as a file to permanently retain an exact duplicate of

About Electronic Publishing and Copyright

the original. The HTML, graphics, and sound files are just as good, quality-wise, as those on the original server from which they were taken.

Consequently, copyrighted material on the Web is often available to anyone who wants it, and copies can spread rapidly with little or no degradation and virtually no control. So consider any technical report you place on a Web server to be public material in a global information infrastructure; and expect that people will look at, copy, and use it freely. Of course, you could encrypt the documents, as we do credit card numbers for on-line transactions, or you could place the materials on a protected server. But these inherent limitations on access may undermine the purpose for having the materials on-line in the first place. And no protection scheme is totally foolproof.

Other possibilities for protecting electronic documents exist, including digital watermarks and various licensing schemes. But when all is said and done, the Web is still a global community—and the rules and values you embrace may not apply everywhere. For copyright protections to be effective, they must be international, specifically applicable to on-line information, and enforced for all users who have access to the information. That may be where we are now headed, especially with the new Digital Millennium Copyright Act (DMCA).[3] But we are not there yet. So think carefully about what you put on-line, and why you are putting it there, before actually doing so.

Checklist for Electronic Publishing

- Have you properly divided the word processing file into subfiles before converting the document to HTML?
- Have you considered user capabilities and loading times when designing your document?

- Have you viewed your document on more than one browser and display screen to check the formatting?
- Have you designed the top level to expose the maximum number of major links?
- Do you have a good reason for publishing this report electronically?
- Have you fully considered the copyright implications of putting your report on the Web?

---

1. Substantial numbers of printed and on-line guides, references, and resources exist for virtually every aspect of electronic publishing. A good general guide to HTML programming and electronic publishing, however, is Dave Raggett (ed.), Jenny Lam, Ian Alexander, and Michael Kniec, *Raggett on HTML4,* 2nd ed. New York: Addison-Wesley, 1998.

2. For more information see Internet: http://www.microsoft.com/technet/.

3. The Digital Millennium Copyright Act (DMCA) addresses several issues raised by digital media, particularly computer networks, and especially the Internet. Among other things, the new law protects system administrators from liability for certain materials that may pass through their systems, better defines what constitutes copyright infringement on computer networks, lengthens the term of protection for copyright holders, and outlaws certain devices designed to circumvent copy and copyright protections.

Notes

17

Ethical Considerations

You may be wondering why a chapter on ethics would be included in a technical writing book. Increasing emphasis is now being placed on ethics in technical writing, as well as in engineering and the sciences. Technology and the communication of technical ideas represent powerful tools for promoting good or evil in society.

Of course, humanity has been struggling for a long time over exactly what constitutes ethical behavior. It is unlikely that this chapter in a technical writing book is going to provide the definitive answer that will end this timeless struggle. The hope, however, is that this chapter will provide some minimal awareness of the ethical dimensions of technical writing and some sensitivity to the real issues that exist within these dimensions. For a more extensive treatment of ethics in the context of engineering and science, many excellent books exist, including Martin and Schinzinger's *Ethics in Engineering*.[1]

What Are Ethics in Technical Writing?

The first thing we need to develop is a working definition of *ethics* in technical writing. To do this, we could struggle with abstract concepts of happiness and goodness and the human condition. We could also talk about what morality is, and we could explore the concept of human virtue. But this kind of extended discussion would be more appropriate for a philosophy book than a technical writing book.

This book's purpose and focus is technical writing—which has, as one of its hallmarks, getting to the point and saying what needs to be said. So we will skip the philosophical issues. What exactly are ethics for a technical writer? For the purposes of this book, the answer is relatively simple:

> *Ethics* are using communication skills and resources with the intention of doing good.

Ethical behavior for technical writers, then, is applying the power of technical communications with the purpose of doing worthy things. But what do good and worthy mean?

You probably already know what you *think* they mean, but not everyone would agree with you. As mentioned in Chapter 16, the global technical writing community, especially with the advent of electronic publishing and the World Wide Web, is composed of many disparate cultures. To some extent, concepts of *good* and *bad,* and *ethical* and *unethical,* are culturally relative. That does not mean we can just ignore the issue, but we need to approach the matter realistically in the context of the pluralistic environment in which we function. That is what this chapter attempts to do.

Begin by considering the following situation:

> A middle-aged technical writer writes an irreverent, humorous technical writing book using, as its examples, reports on fictitious technologies he names after himself. His laboratory report chapter deals with something called the "FinkelTUBE," his research report chapter describes a quantum computing processor known as the "FinkelCHIP," and his proposal and progress report chapters are built around scientific analysis of a figure skating jump called the

"QuadFINKEL." For the sake of argument, assume that the book is a big hit, the royalties pour in, and the author deposits the checks and then goes out and lives the good life.

Now consider this situation:

A middle-aged technical writer writes a professional proposal from a fake corporation he names after himself. He invites individual investment of funds in developing and marketing the fake company's fictitious product line of high-technology devices. He cites invented test data and technical research information demonstrating the efficacy of these products. In fact, he develops a professional, full-color brochure complete with charts and diagrams—and promises huge returns on investment. He mails this brochure to thousands of nursing home residents across the country. When many of these people send him their life savings, he deposits the checks and goes out and lives the good life.

In both situations a technical writer has used his knowledge and resources to develop published materials around fictitious technologies. In both situations he has embellished the capabilities of these technologies with professionally developed diagrams, charts, and data. In both situations he makes money from the published product.

In ethical terms, how do these situations differ? That is an important question; the difference gets at what makes the behavior of technical writers ethical or unethical. Before we get to this difference, however, we need to look briefly at the kinds of ethical constructs that are traditionally used in technical writing and the implications of these constructs for technical writers:

- *Technical writers must be accurate in their work.* Technical writers must be precisely correct at all times, or they are unethical.

- *Technical writers must be honest in their work.* Technical writers who write untruths are unethical.
- *Technical writers must always honor their obligations.* Technical writers who do not produce the documents and other materials they are responsible for producing within the agreed-upon time-frame are unethical.
- *Technical writers must not substitute speculation for fact.* Technical writers who do not clearly separate opinion from accepted truth are unethical.
- *Technical writers must not hide truth with ambiguity.* Technical writers who play down facts that would be contrary to the theses of their reports are unethical.
- *Technical writers must not use the ideas of others without giving proper credit.* Technical writers who fail to document the sources of all nonoriginal ideas, except for common knowledge, are unethical.
- *Technical writers must not violate copyright laws.* Technical writers who fail to document the use of copyrighted materials when used with permission or under "fair use" are unethical. Additionally, technical writers who use any copyrighted materials *without* permission when these uses *are not* covered by "fair use" are unethical whether the materials are documented or not.
- *Technical writers must not lie with statistics.* Technical writers who manipulate data or graphical representations of data, use inappropriate or improper statistical tests, or employ loaded statistical samples are unethical.
- *Technical writers must not inject personal bias into their reports.* Technical writers who are less than objective in everything they write are unethical.

What a list! These rules are clearly well intentioned, and they provide generally useful guidelines for writers. But there is a real problem with ethical rules like these: they miss the mark where ethics are concerned. For example, being accurate and precise in technical writing is not the real ethical issue here. *Intending* to be good and do good is the real issue. Of course, it is true that in technical writing "being good and doing good" usually means being accurate and precise, *but not always.*

In the first situation, as the author in question, I obviously do not believe that being accurate and precise is necessary to achieving my pedagogical goals for this book. In this case, I believe that just the opposite is true. The freedom to invent obviously fictitious technical writing examples supports the book's educational objectives. I am being inaccurate, but I am not being deceptive. After all:

- Do you suppose that anyone really believes that I can do the QuadFINKEL figure skating jump, much less even stand up on a pair of figure skates?
- Do you believe that there's a FinkelTUBE—and if so, that it performs as described?
- Do you suppose anyone believes that a quantum computing FinkelCHIP exists—and if so, that it actually works?
- Or, for that matter, do you think that there is any kind of "Finkel-*anything*" even remotely similar to what has been described in this book?

Of course not!

My point is this: I would be deceptive, and therefore unethical, only if I asked that you, the reader, truly believe these examples. My clear intent is not to mislead; rather, it is to use the fictitious examples to help teach—and I believe teaching is still placed in the general category of "good" things to do. However, I wouldn't do well with that list of rules, would I? These rules do not get at ethics; rather, they get at behavior that is often correlated with ethics.

In the second example, the guy trying to rip off nursing home residents is obviously an unethical toad (he is so bad, in fact, that this statement disparages toads!). He clearly *intends* to use his technical communications skills and resources to deceive. To make matters worse, he is also targeting a vulnerable group of people for this scam—the elderly, who are least able to afford this kind of attack. So unlike the first situation, this guy is falsifying information not to do good, but to do bad. That is why the technical writer in the second example is absolutely unethical (not to mention being guilty of violating numerous local, state, and federal laws).

Figure 17.1 presents the FinkelModel of Ethics in Technical Writing. Perhaps this model may not work much better than the FinkelTUBE or the

Figure 17.1
FinkelModel of
ethics in technical
writing

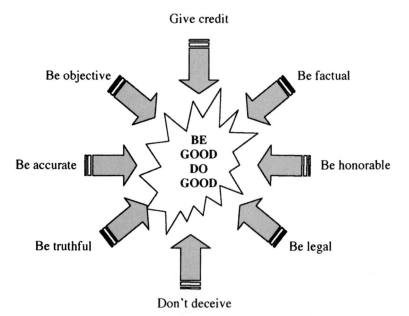

Give credit

Be objective

Be factual

Be accurate

BE
GOOD
DO
GOOD

Be honorable

Be truthful

Be legal

Don't deceive

FinkelCHIP. Clearly it does not represent the big breakthrough in the philosophy of ethics for which humanity has been waiting so long. But it does help make the point and clarify the issue. The rules may contribute to ethics and our understanding of what is involved in ethics, but they are not what constitutes ethics!

Finally, ethics in technical writing, to some extent, must be relative to societal values. If you use technical communications to deceive your reader with the goal of doing bad things (defining *bad* by the standards of your society and perhaps by the standards of civilization as a whole), then you are unethical. If, on the other hand, you use technical communications with the intention of doing *good* things (defining *good* in the same manner), then you are ethical.

---

1. Mike W. Martin and Roland Schinzinger, *Ethics in Engineering,* 3rd ed. New York: The McGraw-Hill Companies, 1996.   **Notes**

18

Abstracts and Summaries

When you have finished putting together your technical report, you still may not be finished writing. You may need to add a special summary of your report, called either an *abstract* or an *executive summary*.

Both abstracts and executive summaries do the same thing: they summarize what is in your report. Abstracts are usually shorter than executive summaries and generally come in two forms: descriptive and informative.

Descriptive abstracts (also called *limited abstracts*) summarize the structure of a report, but not its substance. In other words, descriptive abstracts basically present the table of contents in paragraph form. They refer to the title and the author and may briefly sketch out the purpose, problem, and scope of the document. They also describe the major topics covered by the report. A typical descriptive abstract contains around 50 words.

What Are Descriptive Abstracts?

Writing Descriptive Abstracts

Writing descriptive abstracts is simple because you do not have to get into the substance of the report. Here is a descriptive abstract for the sample research report on the FinkelCHIP in Chapter 10:

Descriptive Abstract

The FinkelCHIP Quantum Central Processing Unit
This research report provides a state-of-the-art investigation and theoretical review of the FinkelCHIP device. The report begins with a description of its purpose, problem, and scope. Next it provides theoretical background on the device's operation. The main discussion deals with the genesis of the FinkelCHIP and provides nonproprietary information on the main parts and basic operation of the device.

Notice how this abstract provides no substantive details about the report except for general topic areas. For example, you know from the abstract that the report discusses the purpose, problem, and scope—but you do not know what the purpose, problem, and scope are. You also know that the report addresses the theoretical background regarding the chip's operation, but the abstract gives no information on this theory. By the way, in some descriptive abstracts, a brief, substantive statement about the purpose and scope may be included.

What Are Informative Abstracts?

Informative abstracts (also called *complete abstracts*) actually summarize the substance of your report, not just the structure. They provide a condensed discussion of the important points. In other words, informative abstracts not only tell the reader the major topics of the report, they also tell the reader, in a nutshell, what you said about those topics. Informative abstracts may or may not be designed as stand-alone documents.

If you intend for the abstract to be part of your report, you need not include the title or author of the report in the abstract. However, if you intend for the abstract to replace the report for some readers, you obviously should include the title (and author, if appropriate) at the beginning of the abstract. In either case, informative abstracts normally contain 100 to 200 words, or less than a single page of double-spaced text.

Writing Informative Abstracts

Writing informative abstracts is a little more involved than producing descriptive abstracts because you need to summarize the substance of the report. That means you need to *understand* the substance of the report. Never try to abstract a report that you do not fully understand. Also, decide early how you are going to organize the abstract. Normally informative abstracts are developed around the main topics of the report. For each of these topics, choose what information to include.

Here is an informative abstract of Chapter 10's FinkelCHIP report:

Informative Abstract

The FinkelCHIP Quantum Central Processing Unit
This research report provides a state-of-the-art investigation and theoretical review of the Finkel-CHIP Quantum Central Processing Unit. Finkel-Products has transcended traditional CPU designs by using quantum nuclear spin states of atoms to store and manipulate large amounts of binary data in parallel. CPU speeds equal to or better than 500 GHz have been realized with this chip. The device has been developed using quantum computing theory to exploit subatomic phenomena of common elements to perform extremely complex computational tasks, resulting in massively parallel processing.

The FinkelCHIP uses nuclear magnetic resonance (NMR) techniques to read specifically induced spin states in carbon, hydrogen, and other atoms. Data are stored and manipulated by using radio frequency (RF) energy to alter the spin states of these atoms while they are trapped in a fixed magnetic field. Different spin states have different energy signatures for different atoms at different magnetic field magnitudes. These differences are read by the NMR sensor, thereby providing an accessible memory for the data stored as spin states. Besides this memory function, manipulating the spin states of atoms also can be used to perform various logic operations.

Building on early quantum CPU successes with carbon and hydrogen atoms, the FinkelProducts' device uses proprietary technology. The FinkelCHIP is built around a Quantum Molecular Matrix, which provides the atomic material; an NMR sensor, which reads the spin states of the atoms; an RF assembly, which provides phased-array RF illumination of individual atoms; and a magnetic field coil, which establishes the required, fixed magnetic field.

Notice the substantive details included in the informative abstract. Also, notice how some materials, such as those related to George Yamaslute and the genesis of the device, have been omitted from the abstract. The fact that Yamaslute demonstrated the feasibility of quantum computing in 1998 was not deemed important enough to be included in the abstract's summary of the report's key points. Writing these kinds of abstracts often requires some tough decisions about what stays and what goes.

What Are Executive Summaries?

An *executive summary* is normally used with large technical reports, such as formal proposals, and other fully developed business or technical documents. Executive summaries are extended, stand-alone abstracts that have both informative and descriptive characteristics. They contain both the substance and the structure of the report. In fact, an executive summary often substitutes for the full report. It is analogous to the class review notes one picks up in the bookstore to avoid buying and reading the course text. Of course, teachers always say that these notes are no substitute for the real book, and they are not—but, as we all know, they do work to some extent.

In a similar vein, executive summaries are designed to provide key management and staff with enough information about what is in a report so that these executives can make informed decisions without reading the entire document. They can always go back and read the document, if warranted—or more likely, have experts on their staff read and analyze the complete report.

Executive summaries can be large documents. Major, formal, multivolume proposals often have executive summaries of 30 pages or more. Also, because these summaries often take the place of the report for key decision makers, these summaries can take on critical importance and must be well written.

Writing Executive Summaries

Writing executive summaries is a challenging undertaking. Your task is to capture as much of the full report's substance as is reasonable, or feasible, in a fraction of the full report's space. To do so, you need a good strategy, especially if the report you are summarizing is complex and extensive. You cannot possibly include everything, so you have to think through what to include and what to leave out of the summary.

A good approach for doing that is to go back to the original purpose of the report. Evaluate everything in the report in terms of (1) how much it contributes to achieving that purpose and (2) how extensive the treatment would have to be if you were to include it. Then include only those portions that contribute the most to the goal of the report and that can be handled effectively in a summary.

As an example of an executive summary, this chapter will summarize this entire book—and, in the process, provide you with a ready-made set of review notes for the book. What is our strategy for boiling down all this material into a few pages? Obviously, everything in this book cannot be included. Some guiding principle for what to include needs to be defined before we start writing the summary. As already mentioned, a good idea is to go back to the *purpose* of the original document. In this case we need to analyze the material in terms of how much it contributes to achieving the goal of this book, which is to explain to engineering and science students how to prepare and edit their technical writing.

Consequently, our strategy is pretty clear: include only the information from each chapter that discusses *how to do something*. That means excluding the FinkelKick, the FinkelCHIP, the FinkelTUBE, and all the other examples. Their function in the book is to complement the "how to" discussions. Only the "how to" discussions need to be in the summary. We will develop this summary around the book's structure, beginning with the first chapter and working through the entire book.

Executive Summary
*Pocket Book of Technical Writing
for Engineers and Scientists*
By Leo Finkelstein, Jr.

The primary purpose of this book is to provide the necessary basics in technical writing so that engineers and scientists can produce and edit technical reports. The book is organized into three major sections (component skills, technical documents, and other considerations) that, together, contain a total of 18 chapters.

Chapter 1: Introduction

Technical writing is a fundamental skill for virtually anyone working in science and engineering. Most science and engineering activities produce technical reports either on paper or in electronic form. Technical writing is the means by which these documents are produced.

Technical writing is not creative writing, which tends to be full of rich metaphors and abstractions. Technical writing is the low-abstraction, high-precision communication of complex technical and business concepts. It is also audience- and situation-relative. The writer must ensure that the reader precisely understands the intended meaning for the purpose at hand. The audience and purpose are almost always well defined for the writer by the technical writing situation.

Technical writing has several unique characteristics that distinguish it from other types of writing. It deals with technical information. It relies heavily on visuals. It uses numerical data to precisely describe quantity and direction. It is accurate and well documented. It uses headings and subheadings for transitions. And it is grammatically and stylistically correct.

Chapter 2: Technical Definition

Virtually any kind of technical writing includes one or more technical definitions. A technical writer must be able to define terms, whether these terms refer to mechanisms or processes. The process of defining involves placing the term into a classification, then differentiating it from other terms in that same classification.

Classifying the term is often the most difficult part of defining it. The class should be a general category in which the term fits, but it cannot be too general. Normally the classification is slightly higher in abstraction than the term itself. Differentiating involves narrowing the meaning of the term to just one possibility within the class.

Chapter 3: Description of a Mechanism

Technology involves mechanisms. Being able to describe these mechanisms precisely and accurately in a way the reader can understand is perhaps the most essential skill of writing technical reports.

Mechanism descriptions are accurate portrayals of material devices with two or more parts that function together to do something. These descriptions focus on the physical characteristics or attributes of a device and its parts. These documents are built around precise descriptions of size, shape, color, finish, texture, and material. Such descriptions also normally include figures, diagrams, or photographs that directly support the text discussion.

The introduction defines the overall mechanism, describes its function and appearance, and lists the parts to be discussed. The discussion section addresses each part by first defining it and then providing detailed descriptions of the part's function and appearance. At the end of each part's discussion is a transition to the next part. The conclusion then summarizes the mechanism's function and relists the parts.

Chapter 4: Description of a Process
Process descriptions are similar to mechanism descriptions, but they focus on the unfolding steps of a process, not physical attributes. Process descriptions can deal with either the operation of mechanisms or the steps of conceptual processes.

The introduction includes a definition of the overall mechanism or conceptual process, descriptions of its purpose and function, and the steps of the process to be discussed. The discussion section first defines each step and then provides detailed descriptions of what happens during the step. At the end of each step's discussion is a transition to the next step. The conclusion then summarizes the mechanism's or process's function and relists the steps.

Chapter 5: Proposals
Proposals are among the most important documents because they obtain grants, contracts, and jobs. Proposals are specialized, technical business documents that offer persuasive solutions to problems. Proposals, unlike other technical documents, need to be more than objective and clear—they need to sell the reader on some idea.

All proposals must do three things: (1) describe, identify, or refer to a problem; (2) offer a viable solution to the problem; and (3) show that the proposing person or organization can effectively implement this solution. Proposals can be formal or informal, and solicited or unsolicited.

Informal proposals include an introduction that specifies the purpose, problem, and scope. They include a discussion section that describes the proposed approach and the benefits that will result from its implementation; they may provide a statement of work that lays out the tasks to be performed. In the resources section, the proposal describes the personnel, facilities, and equipment required to implement the solution. In the costs section, it presents the fiscal and time resources needed to implement the solution. In the conclusion, the proposal summarizes the benefits and risks of adopting the proposed solution and provides a contact for more information.

Chapter 6: Progress Reports

Progress reports, which are also called status reports or milestone reports, follow up accepted proposals by documenting the status of a project. They focus on various tasks that make up the project and analyze the progress that has been made on each task.

Progress reports contain an introduction that covers the purpose, background, and scope of the project. In the status section, the report analyzes tasks that have been completed and provides the status on tasks that are remaining. The conclusion appraises the project's current status, evaluates progress made, forecasts when the project will be completed, and provides a contact for more information.

Chapter 7: Feasibility Reports

Feasibility reports, also called recommendation reports, are documents that recommend the best way to solve a problem. They either determine the feasibility of solving a problem in a particular way or recommend which of several options for solving a problem is the best approach.

The introduction reviews the purpose of the report, the problem that needs to be solved, and the scope of alternatives and criteria. The discussion section is organized around criteria. For each criterion, the report explains the criterion and the data collected and interprets the relative value or effectiveness of each alternative solution based on this criterion. The conclusion summarizes the data and interpretations for all criteria and all candidate solutions, concludes which solution is best, and makes a recommendation based on this conclusion. It also lists a contact for more information.

Chapter 8: Instructions and Manuals

Instructions are process descriptions for human involvement. They not only describe the steps of the process, but they also show someone how to accomplish these steps. Consequently, instructions have to describe a process accurately, and they also have to show a reader how to accomplish each step safely and effectively.

Instructions follow the general format of process descriptions but also include additional material. The introduction is basically the same. The discussion section is different, however. For each step of the process, it does the following: (1) defines the step; (2) gives an overview of what happens in the step; (3) provides needed information specific to the step, such as dangers, cautions, and required equipment; (4) gives specific instructions for accomplishing the step; (5) shows the result that should occur; and (6) provides a transition to the next step (if there is one). The conclusion summarizes the steps of the process and tells the reader where to find additional information.

Chapter 9: Laboratory and Project Reports

Laboratory and project reports present information that relates to the controlled testing of a hypothesis, theory, or device using test equipment (apparatus) and a specified series of steps employed to perform the test (procedure). These reports explain the design and conduct of the test, how the variables were controlled, and what the resulting data show.

Laboratory reports are usually research-oriented documents that start with a hypothesis or theory that needs to be tested. Project reports are often task-oriented documents that start not with a hypothesis but with requirements of a project assignment. Instead of hypothesis validation, these latter reports explain how the tasks were accomplished and assess the success of a project.

The introduction includes the purpose of the report, the hypothesis or requirement that forms the problem, and the scope or limitations of the report. The background reviews relevant theory and past research. The test and evaluation section describes the apparatus and procedure used. The findings section provides the data resulting from the test and interprets these data. The conclusion provides an inference based on these interpretations and a recommendation based on this inference.

Chapter 10: Research Reports

Research reports describe the discovery, analysis, and documentation of knowledge through some type of investigation. They frequently focus on new, evolving, sometimes purely hypothetical technologies, in which case they are called state-of-the-art reports. Research reports are characterized by extensive research and documentation.

The introduction reviews the purpose, problem, and scope of the report. The background section reviews the theoretical basis for understanding the topic and provides a historical perspective of the topic. The discussion presents the main body of research. The conclusion summarizes the material in the discussion and provides recommendations or suggestions based on this summary. The references section includes, at a minimum, sources cited and used. The appendix contains additional supporting material not needed to understand the report.

Chapter 11: Resumes and Interviews

Resumes are specialized proposals by which you offer your services to fill a position. Engineering- and science-related resumes include your name, address, telephone number, and maybe an e-mail address. The objective describes the specific kind of position desired. The strengths section points out your strongest skills and attributes. The education section documents your formal education, including degrees, certifications, honors, and relevant course areas. The computer skills section documents your computer literacy, including operating systems, languages, applications, and platforms. The experience section lists your job experience. The personal section includes job-related personal information that enhances your value to the prospective employer.

Cover letters are used to send resumes to a prospective employer. They should demonstrate your knowledge of the employer and job, briefly summarize your skills and experience, describe your desirable personal traits, and invite a favorable response.

Resumes and cover letters can get you an interview, and an interview can get you a job. When reporting for an interview, bring all the information necessary to fill out a job application, along with extra copies of your resume. Also, fully research the employer before the interview, and be prepared to answer both technical and personal questions. Finally, be prepared to discuss salary requirements, request clarification for interview questions you do not understand, and look professional and act professional at all times.

Chapter 12: Grammar and Style

Grammar is a set of rules providing commonly accepted standards for assembling words so that, together, they make sense and convey meaning. Style is the choice of words and the way we apply the rules of grammar in our writing.

In technical reports, most common grammar and style problems involve the following: comma splices, fused sentences, sentence fragments, misplaced modifiers, passive voice, verb agreement errors, pronoun agreement errors, pronoun reference errors, case errors, and spelling errors.

Chapter 13: Documentation

Proper documentation is an essential element of technical writing—an element that can, and often does, have serious legal, ethical, and credibility implications for those who do not document correctly.

Documentation, in its general meaning, simply refers to creating documents. This chapter focuses on the specific meaning of giving formal credit to a person, organization, or publication for an idea or information that either is not original or is not common knowledge of the field. Documentation is required to meet legal requirements of copyright law, adhere to academic standards, and establish credibility.

Many styles of documentation exist, including notational and parenthetical approaches. Most technical documents use parenthetical references in the text keyed to a list of references at the end of the report.

Chapter 14: Visuals

In communicating complex topics in precise ways, technical writers rely heavily on visuals, which are powerful communication tools that can pack a huge amount of information into a small space. Visuals in technical writing include figures, diagrams, drawings, illustrations, graphs, charts, schematics, maps, photos, and tables.

Use visuals only when you have a specific reason to do so—when they can directly clarify or otherwise enhance your text discussion. Document visuals with source lines, and design your visuals for reproducibility, simplicity, and accuracy.

Chapter 15: Presentations and Briefings

Technical presentations and briefings should be built around substantive ideas and information that are logical, coherent, and well organized. Technical presentations and briefings should also use language appropriate for the audience and be delivered effectively.

Technical speaking situations generally fall into one of three categories: impromptu presentations, in which you have no warning or preparation time; extemporaneous presentations, which are well prepared and rehearsed, but not scripted; and manuscript presentations, in which you read a prepared script. Extemporaneous is the preferred choice.

Technical speaking purposes also generally fall into one of three categories: informative, in which the primary goal is to give an audience facts and other information; demonstrative, in which you show the audience how to do something or how something works; and persuasive, in which you try to convince the audience to make a particular decision or take some specific action.

Technical briefings are focused oral presentations that use visual aids referred to as charts. Charts can take the form of slides, viewgraphs, or computer-generated graphics. A typical technical briefing includes a title chart, overview chart, discussion charts, summary chart, and concluding chart.

Chapter 16: Electronic Publishing

Electronic publishing involves the electronic distribution of printed material and the elaborate networking of information using computers. Technical writing is a big part of this information revolution.

Electronic publishing includes the simple distribution of traditional documents in the form of word processing or document description files. The more complex form of electronic publishing uses hypertext documents encoded with standardized, cross-platform markup languages. Hypertext documents, which can have extensive linking and searching capabilities, are distributed over computer networks, particularly the Internet—as well as via dedicated storage media, particularly CD-ROMs and DVDs.

Electronic documents are similar in content to traditional technical documents, but they are organized into parallel, interrelated structures. This approach differs from the more linear method used in traditional publishing, where paper pages must be assembled in order. Converting traditional documents into hypertext documents involves applying hypertext language tags to the text and reorganizing the material for parallel access.

When designing hypertext documents, take into account the significant data transfer requirements of more complex pages with extensive graphics. Design the top level to show the largest possible number of major links. In addition, take into account the nature of electronic publishing on the Internet, especially where copyright considerations are significant.

Chapter 17: Ethical Considerations

Fundamentally, ethical technical writing means using communication skills and resources with the intention of doing good.

Traditionally, we have approached ethics by laying out sets of rules for what makes an ethical technical writer. These rules have included such things as being accurate and honest, not substituting speculation for fact, not hiding the truth with ambiguity, acknowledging the sources of ideas, complying with copyright laws, not lying with statistics, and not injecting personal bias into reports.

The problem with the traditional approach is that such rules do not get at the essence of ethics: being good and doing good. If you intend to use technical writing to be good and do good, where *good* is defined by your society and culture, then you are acting ethically.

Chapter 18: Abstracts and Summaries

In many cases, technical documents are not complete without a separate section that summarizes the structure or substance of the document. Abstracts and executive summaries provide this function—they summarize the report.

Descriptive abstracts summarize the structure of a report, but not the substance. They refer to the title and author, briefly sketch out the purpose, problem, and scope of the document, and list the major topics of the report. Descriptive abstracts are about 50 words long.

Informative abstracts summarize the substance of a report, not just the structure. They provide a condensed discussion of the important points. Informative abstracts may be designed as stand-alone documents or as a section contained within the report. Informative abstracts are 100–200 words long.

Executive summaries (an example of which you are now reading) are extended, stand-alone abstracts with both informative and descriptive qualities. Executive summaries are used to condense large reports such as formal proposals. They provide key decision makers with enough information to make informed decisions regarding the original report without actually having to read the entire document. Executive summaries vary considerably in length, but they can be quite large. Some often exceed 30 pages where large, complex technical reports are involved.

Index

N

O

P

S

T

U

V